Wild Romanticism

Wild Romanticism consolidates contemporary thinking about conceptions of the wild in British and European Romanticism, clarifying the emergence of wilderness as a cultural, symbolic, and ecological idea.

This volume brings together the work of twelve scholars, who examine representations of wildness in canonical texts such as *Frankenstein*, *Northanger Abbey*, "Kubla Khan," "Expostulation and Reply," and *Childe Harold's Pilgrimage*, as well as lesser-known works by Radcliffe, Clare, Hölderlin, P.B. Shelley, and Hogg. Celebrating the wild provided Romantic-period authors with a way of thinking about nature that resists instrumentalization and anthropocentricism, but writing about wilderness also engaged them in debates about the sublime and picturesque as aesthetic categories, about gender and the cultivation of independence as natural, and about the ability of natural forces to resist categorical or literal enclosure.

This book will be of great interest to students and scholars of Romanticism, environmental literature, environmental history, and the environmental humanities more broadly.

Markus Poetzsch is Associate Professor of English at Wilfrid Laurier University, where he specializes in British Romantic literature and ecocriticism. He is the author of *Visionary Dreariness: Readings in Romanticism's Quotidian Sublime* and has published essays on John Clare, William and Dorothy Wordsworth, Thomas De Quincey, Leigh Hunt, and Henry David Thoreau. His research considers intersecting themes, such as aesthetics and landscape gardening, pedestrianism and loco-description, anthropocentrism and ornithology, poetics, and ethics.

Cassandra Falke is Professor of English Literature at UiT—The Arctic University of Norway. Her books include *Phenomenology and the Broken Body* (co-ed. 2019), *The Phenomenology of Love and Reading* (2016), *Literature by the Working Class: English Autobiography, 1820–1848* (2013), and *Intersections in Christianity and Critical Theory* (ed. 2010). She has published essays on romanticism, phenomenology, education, and the role of the reader. Her current project discusses acts of reading in light of recent theorizations of complicity.

"*Wild Romanticism* is an innovative and highly original collection of essays that makes a substantial and persuasive contribution to the discipline of environmental humanities. The topic of wilderness during the Romantic period is an important and largely unexplored area of scholarship, one that will be of compelling interest to scholars of British and European literature and environmental history. This book will appeal to a broad range of readers due to its bold originality and its relevance to contemporary environmental concerns."

— *James C. McKusick, University of Missouri-Kansas City, author of* Green Writing: Romanticism and Ecology *and co-editor of* Literature and Nature: Four Centuries of Nature Writing.

Routledge Environmental Literature, Culture and Media

Series editor: Thomas Bristow

The urgency of the next great extinction impels us to evaluate environmental crises as sociogenic. Critiques of culture have a lot to contribute to the endeavour to remedy crises of culture, drawing from scientific knowledge but adding to it arguments about agency, community, language, technology and artistic expression. This series aims to bring to consciousness potentialities that have emerged within a distinct historical situation and to underscore our actions as emergent within a complex dialectic among the living world.

It is our understanding that studies in literature, culture and media can add depth and sensitivity to the way we frame crises; clarifying how culture is pervasive and integral to human and non-human lives as it is the medium of lived experience. We seek exciting studies of more-than-human entanglements and impersonal ontological infrastructures, slow and public media, and the structuring of interpretation. We seek interdisciplinary frameworks for considering solutions to crises, addressing ambiguous and protracted states such as solastalgia, anthropocene anxiety, and climate grief and denialism. We seek scholars who are thinking through decolonization and epistemic justice for our environmental futures. We seek sensitivity to iterability, exchange and interpretation as wrought, performative acts.

Routledge Environmental Literature, Culture and Media provides accessible material to broad audiences, including academic monographs and anthologies, fictocriticism and studies of creative practices. We invite you to contribute to innovative scholarship and interdisciplinary inquiries into the interactive production of meaning sensitive to the affective circuits we move through as experiencing beings.

Mediating Nature
The Role of Technology in Ecological Literacy
Edited by Sidney I. Dobrin and Sean Morey

Shakespeare and the Evolution of the Human Umwelt
Adapt, Interpret, Mutate
Timothy Ryan Day

Wild Romanticism
Edited by Markus Poetzsch and Cassandra Falke

For further information about this series, please visit: https://www.routledge.com/Routledge-Environmental-Literature-Culture-and-Media/book-series/RELCM

Wild Romanticism

Edited by
Markus Poetzsch and
Cassandra Falke

Routledge
Taylor & Francis Group
LONDON AND NEW YORK

earthscan
from Routledge

First published 2021
by Routledge
2 Park Square, Milton Park, Abingdon, Oxon OX14 4RN

and by Routledge
52 Vanderbilt Avenue, New York, NY 10017

Routledge is an imprint of the Taylor & Francis Group, an informa business

British Library Cataloguing-in-Publication Data
A catalogue record for this book is available from the British Library

Library of Congress Cataloging-in-Publication Data
A catalog record has been requested for this book

ISBN: 978-0-367-49672-2 (hbk)
ISBN: 978-0-367-75351-1 (pbk)
ISBN: 978-0-367-49674-6 (ebk)

Typeset in Times New Roman
by codeMantra

Contents

Figures

Contributors

Colin Carman is Assistant Professor of English at Colorado Mesa University in Grand Junction, Colorado, and author of *The Radical Ecology of the Shelleys: Eros and Environment* (Routledge Books, 2019). A former fellow at the Huntington Library in San Marino, California, he has contributed to three book collections: *Lacan and Romanticism*, *Romantic Ecocriticism: Origins and Legacies*, and *The Brokeback Book*. His articles, ranging from the Shelleys and Walter Scott to the films of Werner Herzog and Terrence Malick, have appeared in such journals as *ISLE, European Romantic Review*, *GLQ*, *Studies in Scottish Literature*, and *Horror Studies.*

Tristanne Connolly is Associate Professor in the English Department at St. Jerome's University in the University of Waterloo. She is the author of *William Blake and the Body* (2002), along with several articles on Blake, on Erasmus Darwin, and on British Romantic literature in relation to science and medicine, gender and sexuality, and religion. She has edited a number of essay collections, most recently *Beastly Blake* (2018), with Helen P. Bruder, and *British Romanticism in European Perspective*, with Steve Clark (2015).

William Davis is Professor of Comparative Literature and German Studies at Colorado College and works on intersections between philosophy and literature, comparative British and German Romanticisms, and literary theory. He is the author of *Romanticism, Hellenism, and the Philosophy of Nature* (2018). His articles on German and British Romanticism, often viewed in relation to German Idealism, have appeared in journals such as *Goethe Yearbook*, *Germanic Review*, *German Quarterly*, *European Romantic Review*, *Prisms*, and *The Wordsworth Circle*. He is currently working on a project on material culture and the invention of the "classical" in the late eighteenth and early nineteenth centuries.

Cian Duffy is Professor and Chair of English literature at Lund University, Sweden. He has published monographs, editions, and articles dealing with various aspects of the intellectual life and cultural history of Europe during the Romantic period. Particular focal points have been representations of landscape and the work of Percy Bysshe Shelley. Recent work includes the collection *Romantic Norths: Anglo Nordic Exchanges, 1770–1842* (Palgrave, 2017) and the edition *Percy Bysshe Shelley: Selected Poems and Prose* (ed. with Jack Donovan; Penguin, 2018).

Sue Edney is a Senior Associate at Bristol University, specializing in Romantic and Victorian poetry, language, and ecocriticism. She has published on dialect, identity and place, William Barnes, Tennyson, Philip Henry Gosse, Gothic sea-anemones, and environmental justice, in addition to book reviews. She is co-editing two essay collections: *Reworking Georgic*, with Tess Somervell, and *Hannah More in Context*, with Kerri Andrews. She is Reviews Editor for *Green Letters: Studies in Ecocriticism,* the journal for ASLE-UKI, and the International Ecolinguistics Association steering committee's ecocriticism representative. Her essay collection *EcoGothic Gardens in the Long Nineteenth Century: Phantoms, Fantasy and Uncanny Flowers* (MUP) is forthcoming November 2020.

Cassandra Falke is Professor of English Literature at UiT—The Arctic University of Norway. Her books include *Phenomenology and the Broken Body* (co-ed. 2019), *The Phenomenology of Love and Reading* (2016), *Literature by the Working Class: English Autobiography, 1820–1848* (2013), and *Intersections in Christianity and Critical Theory* (ed. 2010). She has published essays on romanticism, phenomenology, education, and the role of the reader. Her current project discusses acts of reading in light of recent theorizations of complicity.

Mirka Horová is Senior Lecturer in English at Charles University and Editor of *The Byron Journal.* She has written widely on Byron, including chapters on heroic transformation (Routledge, 2016), Italian dramas (MUP, 2017), Lucretius (CSP, 2018), the Satanic School (CUP, 2019), and an article on swimming (2019). She has edited several books and special journal issues. From 2013–19, she co-organized the international Newstead Abbey Byron conference. Her research interests include Romantic literature and play theory. This work was supported by the European Regional Development Fund-Project "Creativity and Adaptability as Conditions of the Success of Europe in an Interrelated World" (No. CZ. 02.1.01/0.0/0.0/16_019/0000734).

Gregory Leadbetter is Professor of Poetry at Birmingham City University. His research focusses on Romantic poetry and thought, and the traditions to which these relate, together with the history and practice of poetry more generally. His book *Coleridge and the Daemonic Imagination* (Palgrave Macmillan, 2011) won the University English Book Prize 2012. His poetry collections include *Maskwork* (Nine Arches Press, 2020), *The Fetch* (Nine Arches Press, 2016), and the pamphlet *The Body in the Well* (HappenStance Press, 2007).

James Lesslie was awarded a PhD from Birkbeck College, University of London, in April 2020 for his thesis "British Women Writers and the Culture of Wild Nature: 1781–1815." His research interests include women's writing in late eighteenth- and early nineteenth-century Britain—particularly representations of the natural world and their interconnections with revolutionary-era political and ideological debates—as well as historiography, travel writing, Gothic literature, and regional cultures and identities in Britain during the long eighteenth century.

Emma Mason is Professor of English and Comparative Literary Studies at the University of Warwick. Recent publications include *Reading the Abrahamic Faiths: Rethinking Religion and Literature* (Bloomsbury, 2015) and *Christina Rossetti: Poetry, Ecology, Faith* (Oxford University Press, 2018). She is series editor of Bloomsbury's monograph series, New Directions in Religion and Literature, with Mark Knight, with whom she is currently writing *Weird Faith in Nineteenth-Century Literature: Theologies at Work.*

Markus Poetzsch is Associate Professor of English at Wilfrid Laurier University, where he specializes in British Romantic literature and ecocriticism. He is the author of *Visionary Dreariness: Readings in Romanticism's Quotidian Sublime* and has published essays on John Clare, William and Dorothy Wordsworth, Thomas De Quincey, Leigh Hunt, and Henry David Thoreau. His research considers intersecting themes, such as aesthetics and landscape gardening, pedestrianism and loco-description, anthropocentrism and ornithology, poetics, and ethics.

Robert W. Rix is Associate Professor at the University of Copenhagen. He has published widely in several areas relating to the eighteenth and nineteenth centuries: politics, religion, language, nationalism, Nordic antiquarianism, and print culture/book history. In a number of articles and a monograph, Rix has focussed on William Blake and the religious radicalism of the 1790s. He has also written on medieval ideas and manuscript culture in *The Barbarian North in Medieval Imagination: Ethnicity, Legend, and Literature* (2014). Rix's forthcoming book examines the exploration and representation of Greenland in Western culture and literary works.

Acknowledgements

In normal times, a volume such as this depends on the collective inspiration, synergy, diligence, patience, generosity, and goodwill of its contributors, without whose application there can be no success. Against the backdrop of COVID-19, when most of the writing and editing took place, all of these admirable qualities have had to be quickened and intensified—and a few more virtues thrown into the mix! Some of our contributors dealt with personal and family illness; some worked in the midst of a frantic transition from in-class to online teaching; some scrambled to find texts and sources in a time of shuttered libraries; some (okay, most) struggled with periods of profound stress, isolation, and uncertainty. And yet somehow, despite the odds, the work was completed on time and in a manner consistent with our hopes. We thank you.

Our gratitude must also go to the three anonymous reviewers on whose advice Routledge sped this book to print. We are indebted to you for your astute commentaries, helpful suggestions, and enthusiastic support. In like manner, we thank our Routledge team—Oindrila Bose, Rebecca Brennan, and Tom Bristow—for their belief in the project and their invaluable assistance throughout the planning and production phases.

Cassandra Falke would like to thank her husband, Damon, whose support in a period of illness made work on this book possible, and her two sons, Charlie and Sebastian. I am thankful for our years of exploring the wild together. This volume grew out of the Wild Romanticism conference in Tromsø, Norway, in 2018. Thanks are due to UiT for their support of that event and to conference participants for their inspiring presentations and spirit of adventure.

Markus Poetzsch would like to thank his wife, Jeni, and his sons, Sam and Gabriel, for those "little, nameless, unremembered acts of kindness and of love," against which even a pandemic cannot prevail.

Introduction

Cassandra Falke and Markus Poetzsch

> The progressive reinvention of wilderness as something to be treasured—'unspoiled nature'—paralleled the increasing urbanization and industrialization of Western society . . . Taking the lead in articulating and promoting this new view of wilderness was a group of Romantic writers.
>
> (Charles Warren, "Wildness," International Encyclopedia of Human Geography 281–282)

> To the extent that wilderness spaces and the laws that created them persist, we are still living, literally, within the Romantic period.
>
> (Timothy Morton, *Ecology without Nature* 114)

Although Romanticism frequently appears in Wilderness Studies, and wilderness appears in studies of Romanticism, the relationship between the two is often assumed rather than directly examined. Romanticism is related to certain concepts of wilderness both historically and as a trans-historical phenomenon of perception. The association between Romanticism and wilderness seems like a matter of history, as geographer Charles Warren frames it in the quote above. Alongside the industrial revolution, a group of Romantic writers—"urban-based literary" types—perceived the wild as an "appealing other" that could summon forth some element of selfhood kept hidden in the throng of cities, and along with "proto-conservationists," "landscape artists and photographers," this group propagated a concept of wilderness that was anti-urban, and anti-industry, something that could be conserved as an act of resistance to modernization (281–282). In this view, Romanticism is important to our understanding of wilderness as a point of origin for a wilderness concept that has led to policies protecting of specific wild spaces, like national parks and wildlife sanctuaries. Timothy Morton, in contrast, sees wilderness as perpetuating a certain form of Romantic subjectivity. Wilderness relies on "social and psychological" distance. The Romantic subject portrays him- or herself as simultaneously within a perceptual field characterized by distance and at the edge of that perceptual field observing the self—the overcoated, stick-toting figure of Caspar David Friedrich's *Der Wanderer über dem Nebelmeer* (*Wanderer above the Sea of Fog*) and the painter admiring him all in one. Wilderness, according to Morton, only becomes recognizable as such when one is outside of it. In this view, wilderness and Romanticism are so intimately related that the one perpetuates the other.

Wild Romanticism approaches the relationship between wilderness and Romanticism both as a matter of history and as a matter of conceptual co-creation, while challenging the idea that the wild is something history has moved beyond, or which persists in the present only as nostalgia. The collection finds the wild to be an actively renewing force that is associated with distance because of its resistance to human control, not because it is spatially far away or temporally left behind. The "wild" conjures up notions of what is culturally, politically, and cognitively unassimilable. *Wild Romanticism* explores wildness as a trait that Romantic authors yearned towards and directed attention to, a trait that they attributed to spaces we would still be likely to call "wilderness," but that they also found close to home. For Romantic authors, "wild" describes internal states (wild imagination) and external manifestations (wild eyes). Plants, people, animals, and landforms can all be wild. The wild goes in and out of houses, riding along in hair and glances, and yet resists any form of domestication. The nuance and adaptability of Romantic concepts of wildness have been lost, somewhat ironically, in the broad recognition of Romanticism's importance to later conceptions of wilderness.

Etymologies of the wild

During the nineteenth and twentieth centuries, the meanings of "wilderness" and its cognate "bewilder" changed. "Bewilder" came to be associated with internal states and "wilderness" with external spaces. This is due, to a large extent, to what Jedediah Purdy calls "a new public language of wilderness" that developed in the mid-twentieth century (190). The "Wilderness Movement" began in the US in the 1920s as conservationists became divided over the purpose of national parks. One group advocated a consumer model of parks usage that focussed on tourists' access to beautiful scenery and outdoor recreation, while another focussed on the preservation of land unaltered by human interference. By 1964, the Wilderness Preservation Act defined such land as "wilderness." The wild was an "area where the earth and its community of life are untrammeled by man, where man himself is a visitor who does not remain" (US DoJ). This concept of the wild as "undisturbed by human activity" has secured such a firm place in international law (IUCN) and cultural parlance that it is difficult to recover the flexibility and interrelatedness that wild/wildness/wilderness retained in the Romantic period. In the early nineteenth century, the wild was not exclusively, or even primarily, defined based on its separation from human activity. John Keats's work exemplifies this well. Noah Comet's concordance of his work shows that Keats uses the term "wild" seventy-one times in his collected works, twenty-two times in *Endymion* alone. Keats describes explorers' "wild surmises" ("On First Looking into Chapman's Homer" line 13), and a young man's "wild uncertainty" (*Endymion* II: line 273). The act of "conjecturing" can be "wild" (*Endymion* III: line 556), but so can a "forest nook" (*Endymion* II: line 890). The first three of these examples refer to individual or communal acts of thought as wild, suggesting that like a forest nook, a thought can appear that was unlooked for, wondrous, and potentially dangerous, without prior intention or after-the-fact subsumption into existing categories. "Wild" thoughts include surmise, uncertainty, and conjecture, and therefore retain some of the flexibility Keats associates with creative, negatively capable thinking.

"Bewilder" is a related term, which etymologists think is a back formation derived from wilderness (Hoad). According to Google's NGram viewer, the word reached its peak of popularity exactly during the Romantic period, between 1776 and 1832. Seventeenth-century uses of the term retain the theological weight of the wilderness being the space outside the Garden of Eden, as in John Scott's *The Christian Life*, which despairs of humanity as "almost universally lost and bewildered" (3: 101), but by the Romantic period, postlapsarian associations had faded. *The London Encyclopedia* of 1829 defines "bewilder" as "to lose in pathless places; to confound for want of a plain road; to perplex; to entangle; to puzzle." Like its cognate "wild," the term "bewilder" mediates internal and external states, but it emphasizes in both cases an enforced directionlessness. Turning to Keats again, we find "bewildered shepherds" hoping to regain their path quite literally (*Endymion* 1: line 269), but also a "brain bewilder'd" ("To My Brother George" line 2). According to the *Oxford English Dictionary*, the association between bewilderment and literal pathlessness became obscure after the mid-nineteenth century, but for Romantic-period authors, the experience of wandering in physically uncharted space seems to have lingered in the experience of psychological bewilderment. This residue of literal pathlessness hints at crucial differences between the early nineteenth and early twenty-first centuries—many people today go years at a time without ever being lost in pathless woods, due to the predominance of urban living and technologies like cell phones and GPS tracking. The early nineteenth-century use of "bewilder" also hints at an element of Romantic wildness that gets overlooked if we link "wild" too readily to aesthetics of sublimity or distance. The vitality of the metaphor behind "bewildered" implies that being literally wilded or wildered was something everyone had experienced, not only explorers or long-trekking poets.

Twentieth-century ideological battles over what defines wilderness have led to the term being associated with a nature/culture divide that no longer seems tenable. Consequently, Warren anticipates twenty-first-century discussions of ecology turning increasingly to "wildness" instead of "wilderness" (288). *Wild Romanticism* acknowledges, even celebrates, the important role that Romantic authors had in articulating ideas that became central for wilderness conservation, but in choosing the more dynamic "wild" over "wilderness" for our title, we want to reclaim some of the term's nineteenth-century slipperiness and to highlight the resonances between the experience of existential or cognitive be-wilderment and the embodied experience of being directionless, out of control. Morton seems surprised that the "space of [John Clare's] village, even if it was indeed feudal, was always already criss-crossed with otherness" (200). He writes that Clare's poem "I Am" is "the stunning moment at which this otherness is perceived as intrinsic to the self" and that in this same moment "we have lost nature, but gained ecology" (200). Morton associates "nature" with a conceptual covering up and homogenizing of otherness, a projection of such otherness away from the self. "Ecology," in his usage, bespeaks a willingness to relinquish the desire for conceptual mastery and a recognition that human selves exist within and contribute to global and local environments that include non-human species, landforms, and geologic processes. Clare becomes ecological when he realizes that an unnameable otherness pervades every place, even those that seem most familiar. Ecology knows that it does not know; nature names and claims. But in rejecting nature in favour of ecology, Morton is looking for something the wild seems to have already provided for Romantic-period authors, a way of marking the unknowable that originates from outside human subjectivity and remains resistant to it even when we experience it within our own body or consciousness.

The romantic wild

Within Romantic ecocriticism, by which we mean both ecocriticism influenced by Romantic-period ideas and a subset of that ecocriticism devoted to Romantic-period literature, there has been a tendency to trouble the divisions between nature and culture, human and non-human, the country and the city. The focus on these conceptual dichotomies has led contemporary scholars to overlook wildness, which many Romantic-period authors use to mediate these divisions. This collection seeks to fill a gap in Romantic scholarship by consolidating thinking about Romantic conceptions of the wild, clarifying the emergence of wilderness as a cultural, symbolic, and ecological idea. This gap is particularly apparent in studies of Anglo-European Romanticism. As Phillip and April Vannini point out, "most volumes on wilderness are in one way or another about the US and about American ideas, histories, values and perspectives of wilderness" (22). Important theorizations of Romantic ecology by scholars such as Jonathan Bate, Karl Kroeber, Kate Rigby, and Timothy Morton, which have focussed on Britain and Europe, have not made wildness or wilderness a central focus of their work in the way that scholars of nineteenth-century America have. Yet as James McKusick points out, for Anglo-European

Romantic authors, the "natural world is pervaded by revolutionary energies" that express themselves in what is unruly, ungovernable, and unfathomable ("Introduction" para. 4).

Celebrating the wild provided Romantic-period authors with a way of thinking about nature that resists instrumentalization and anthropocentricism while also engaging them in debates about the sublime and picturesque as aesthetic categories, about gender and the cultivation of independence as natural, and about the ability of natural forces to resist categorical or literal enclosure. Scholars included here examine spaces like the Arctic, which still sometimes functions metonymically as the boundary of human power, but they also discuss sites where wildness shows up unexpectedly, like William Blake's garden and Jane Austen's parlours. They deepen knowledge about authors who appear frequently in Romantic ecocriticism, such as William Wordsworth and John Clare, but also explore work by Ann Radcliffe, Friedrich Hölderlin, and James Hogg, whose writings about the wild have been neglected. Also included are Lord Byron and Samuel Taylor Coleridge, who used the term "wild" more freely than most, but who do not appear frequently in ecocritical scholarship. There is still more work to do to expand knowledge about nineteenth-century conceptions of the wild, particularly in European Romanticism, but we have tried to curate the collection to reflect diverse geographies and authors. The collection covers wild spaces from Sicily to Greenland, from Blake's back garden to Mary Shelley's fields of ice.

Taken together, the essays here reveal the Romantic wild to be more pervasive than previously imagined. The wild is characterized in terms of otherness and uncontrollability without necessarily being projected into a distant or uninhabited place. Percy Shelley writes famously about the steep-sided precipices of Mont Blanc, but the wild also pervades his love poems to Jane Williams. The illicit lover does not picture himself in relation to a sublime edifice, but to wilting flowers and suffering pumpkin plants. The wild and numinous tone that pervades landscape descriptions in "Kubla Khan" recurs throughout Coleridge's writing in references to his own armchair-thinking processes as "wild," particularly when he tries to push his mind beyond what has been thought before. The wild, for authors examined here, is not something to be conquered or mapped, but something that we exist within and among. The wild that Hogg's character Allan Gordon discovers in Greenland, for example, surrounds a small village and invades the village in the form of polar bears. The humans in the story are not portrayed as masters in the land they inhabit, but rather exist precariously alongside other animals struggling for survival. Clare, whose fenny home landscape had been worked and reworked, writes of the plenitude of wild life in the fields, rather than longing for some imagined, never inhabited wild. Positioning himself below the birds, among their nests, within a copse, beside a mouse, he repeatedly testifies to the surprising activities and non-human intentionality of species around him.

A paragraph from Wordsworth's *Guide to the Lakes* reveals the pervasiveness and flexibility of "wild" terminology. Wordsworth is pronouncing upon the proper way of decorating houses ("dust colour" is acceptable, but white or yellow is not; "foreign" fauna may only be grown near the door), but the problem of larch plantations, which he discusses at some length, leads him to the topic of artificial versus natural tree and shrub propagation (para. 85–86). He suggests that the space between a house and a forest be filled with shrubs and fruit trees found in the woods, and here "wild" and "wilding" serve as species markers to distinguish "the wild-rose," "the wilding, black cherry tree," and the "wild cluster-cherry" or hackberry from their domesticated cousins. These plants retain a wild identity although they survive transplantation. Wordsworth therefore deems them appropriate to mediate domesticated and non-domesticated spaces. Wildness here is a continuum that a human designer can respond to and position him or herself within, but that it would be foolish to try to control or fully emulate. Later in the paragraph, Wordsworth pairs "wild" and "bewildered" in a way that reveals the concept's ambiguity. He contrasts a "wild wood," admirable in its naturalness, with a cultivated space "where we have the whole contents of the nurseryman's catalogue jumbled together—colour at war with colour, and form with form." There, in the designed portion of the property, he finds "among the most peaceful subjects of Nature's kingdom, everywhere discord, distraction, and bewilderment!" (para. 87). The "peaceful subjects" seem to be the trees and shrubs themselves, and the source of bewilderment among them the "artificial planter," a human (para. 88). Everything in this paragraph seems to have agency. Forcibly juxtaposed colours war with one another, while plants crave peace. Cherry trees wild (a verb), and humans bewilder. Humans are decentred but not absent in this consideration of what is wild and what wilds. They become a wilding or bewildering force that threatens the peaceful tree subjects in something of the same way Hogg's polar bears threaten human Greenlanders.

Wordsworth's emphasis is not on the wild as something beyond a boundary of human habitation, but beyond human domination. The enclosures of small farmers produce "wild graces" (para. 93), and cottages added to over time possess "wildness and beauty" (para. 66). "Artificial planters" reveal bad taste in behaving as though the growth of trees and production of colour were within their control. In contrast, on the mountainside, no force dominates another. The forces of trees protecting one another are balanced by the forces of soil deterioration and wind. Humans would do better, Wordsworth suggests, to receive and interpret the direction of multiple natural forces, to take a responsive rather than a controlling role, much as the trees themselves do. The first settlers, Wordsworth imagines, followed with their ploughs: "the veins of richer, dryer, or less stony soil; and thus it has shaped out an intermixture of wood and lawn, with a grace and wildness which it would have been impossible for the hand of studied art to produce" (para. 47). In all these instances there is an emphasis on process and passive reception of forces beyond human agency.

Ashton Nichols connects "the Romantic idea of wilderness" to the image of "a space so vast that the edges of it can hardly be conceived, as a mysterious wildness with which humans have almost nothing to do" (xvi), but looking at the *Guide to the Lakes*, we find something very different. Wildness is mysterious in small as well as large manifestations. The wild cannot be conceived, but not because humans have nothing to do with it. Rather, the wild marks the boundary of human control, both physically and conceptually. One can work with the wild—steering a plough along a seam of more yielding ground or moving a wild rose into the garden—and if the wildness of the rose or the dirt resists manipulation enough, then we can trace the boundary of where the wild begins, but the human role in this process is allowing wildness to manifest itself, not conceptually subsuming it.

External geographies

The Romantic Wild, as this volume makes clear, has many faces (and perhaps as many masks), yet its primary orientation is towards what is other, what lies outside the authority and security of the egoic subject and manifests itself in "things we have not created ourselves" (McKusick 10). For Romantic-era writers, wild "things" were almost invariably natural things and their thingliness was distributed across spaces of diverse configuration, varying scope, and inevitable alteration, but all of them related by the workings of a power that resists regulation. The common names for these spaces, readily evident in the titles of travel literature, poetry, fiction, and aesthetic treatises of the period, are of course familiar to us still and reveal something about the disciplines brought to bear on their description and the tactics deployed in their management: spot, scene, prospect, land, island, landscape, topography, region, county, country, world. Authors who evoked these terms, whose work we now divide tidily by genre or discipline, were all doing a form of geography, literally a writing of the earth. Yet in the eighteenth century this academic discipline—still in its "'pre-critical' stage of development" (Mayhew 385)—directed its enquiries rather far afield and thus often passed over the smaller or more local manifestations of wild space, and this tendency to overlook the local has persisted in geographical studies of wildness until quite recently. As Robert Mayhew points out, physical geography in the Romantic period was "centred on acquiring knowledge about the situation of places in terms of their longitude and latitude" (390). Samuel Johnson accordingly establishes the geographer's proper domain, in a definition drawn from the tracts of Isaac Watts, as "the knowledge of the circles of the earthly globe, and the situation of the various parts of the earth." In such a geography there is no place for the heaths and woods of

Helpston, the arbour at Nether Stowey, the garden plots of Dove Cottage or, alternately, Lambeth, the plum-tree at Wentworth Place, or the upper-floor apartment of potted plants at Tre Palazzi di Chiesa.

And even the larger expanses of geographic space to which Johnson's definition confidently alludes were, as Michael Wiley reminds us, but imprecisely and inadequately mapped. "Through the end of the eighteenth century, geographers did not know the full exterior boundaries of even the world's major land masses—areas seemingly accessible to the scientific eye" (11). Selecting one notable example, Wiley points out that Romantic-era cartographers did not conceive that Greenland is separate from North America. And even where apparently familiar and well-travelled regions such as the West Indies were concerned, many eighteenth- and early nineteenth-century maps betray an abundance of "empty spaces" (Wiley 11) outside of the British colonies themselves. Thus, alongside the well-established and exhaustively documented narrative of European colonization in this period—with "the known world," as Katherine Turner avers, "expand[ing] at a prodigious rate" (22) and an astonishing twenty-six per cent of the global population under British rule by 1820 (Leask 273)—runs the thread of another more complicated story about the limits and lacunae of geographic knowledge, what it seizes but fails to grasp.

The tension between the known world and one that operates at the margins or wholly outside the structures of Anglo-European epistemology is important to the purpose of this volume because "empty spaces" have historically (and indeed still today) been most readily populated by our notions of the wild. That is where the energy of wildness runs first, into the breaches of cartographic certainty, into the lapses of colonial order and control, into the recesses of the unknown. Romantic writers, as we will suggest, also of course apprehended and encountered wildness in more domesticated spaces, even indeed in cities—what Gavin Van Horn designates as "the relative wild" (4)—but in contemplating the macro-geography of this period, one cannot simply pass over the fact that their "known world," the product of travel and speculation, literature and lived experience, surveys and maps, was significantly different and inarguably more circumscribed than our own. This is a circumstance that contemporary debates over the meaning or validity of wild spaces, especially those freighted with the contested designation of "wilderness," consistently overlook. Indeed, the constructivist objections that have been raised in regard to the geographic reality or given-ness of wilderness as a pristine landscape remote from the claims and incursions of humanity tend to adopt a selective and often narrow historicism to which is sutured, as David Kidner argues, "a delusory anthropocentric arrogance" (13) that overestimates the human role in shaping the world of nature. William Cronon's *Uncommon Ground* (1996), arguably the ur-text of the modern constructivist position, renders not only wilderness but nature itself as "a profoundly human construction" ("Introduction" 25). What he means of course is not to deny the reality, geographic or otherwise, of our physical environment but rather to interrogate certain symbolic accretions to place—to "empty space," we might say—that have over time filled that emptiness "with some of the deepest core values of the culture that created and idealized it" ("Trouble with Wilderness" 73). Tellingly, in his historical survey of biblical, Romantic, nineteenth- and twentieth-century American conceptions of wilderness—these being the only reference points deemed relevant—Cronon makes no attempt to distinguish the geographic realities or "known worlds" of these disparate cultures, peoples, and periods.

More recent forays into the constructivist debate have acknowledged that the modern world of wilderness spoliation and erasure—if that is indeed our only reality now—is not a secure lens through which to study the past. Emma Marris's *Rambunctious Garden: Saving Nature in a Post-Wild World* (2013), for example, though it also dispenses with "our Romantic notion of untrammeled wilderness" (2) and dallies (as its subtitle suggests) with "Anthropocene boosterism" (Butler xi), concedes as a necessary foundation for analysis that "We have lost a lot of nature in the past three hundred years" (1). To put it simply, Romantic-era writers and travellers had more reason to have faith not only in the construct but also in "the *actuality* of wilderness" (Nash, "Wild World" 186, emphasis added) than many people have today. In making this claim, however, we do not deny that their representations of wilderness were often, like our own, given to fanciful reconstruction or outright invention. As Christoph Bode and Jacqueline Labbe suggest, the focus of writers in this era is on "actual locations . . . imaginatively conceived" (3). This distinction is important because it allows us to examine not only the "empty spaces" of the Romantic geographic imaginary but also, at the same time, the nature of its "known world" and how it made space for wild space.

A return to Johnson's definition of geography reveals the easy slippage from an outward-looking, ostensibly fact-based and scientifically corroborated body of knowledge (e.g. cartography) to something rather more introspective, contingent, and politicized. As he notes, "in the largest sense of all, [geography] extends to the various customs, habits, and governments of nations." This mapping of what we now designate as human geography or anthropogeography onto the cartographer's parchment is precisely what accounts for the prominence of colonial settlements and their surrounding "empty spaces" on Romantic-era maps of the West Indies. Not only were maps of the "known world" ideologically inflected (and indeed they still are, as the differences between the Mercator and Gall-Peters projections reveal), but so were the travel narratives that served to animate wild as well as cultivated space for eighteenth- and nineteenth-century readers. These texts, as Nigel Leask points out, had tremendous influence on the mental geography of their readers, providing both rural and metropolitan audiences "with their knowledge of the 'wider world'" (272). Yet this knowledge was by no means a stable or uniform resource and, as such, neither were Anglo-European impressions of the "wider world." The work of Julia Kuehn and Paul Smethurst, for example, foregrounds the "melding of fact and fiction" in Romantic-era travel literature, isolating the role of mobility as both a potentiating and destabilizing force in the narratives of empire (1–2). Steve Clark, in like manner, finds "The dividing line between fact and fiction" in such writings to be "traditionally elusive" (2), with the implication that consumers of such works—an audience second in number only to readers of novels and romances (Leask 273)—were typically armed with scepticism as well as expectation in confronting these texts. Indeed, we might say that for many eighteenth- and nineteenth-century readers, knowledge of the wider world was not based on a strict adjudication of fact and fiction, a sifting of one from the other, so much as an acceptance of both as an indelible part of the process of making geographic space.

We have adopted a similar approach to an understanding and appreciation of the great varieties of wild space—even wilderness—on offer in an Anglo-European Romantic context. Rather than cleave to the constructivist extreme, whereby nature is evacuated of all presence other than the symbolic, or embrace an essentialism through which the inherently creative process of "making space" discursively is elided, the essays that follow position wildness alternately as "actuality" (to use Nash's term), subjective experience, mutable process, contested construct, and imagined possibility. We concur, in short, with the Vanninis's artful distillation of the issue: "Wilderness may be an illusion, but it is also real in many ways" (31). The Romantics certainly found it to be both, whether in faraway geographies, some pieced together from books alone, or in the immediate contexts of their lived experience—in paths walked, gardens worked, worlds wilded.

Internal geographies

One cannot map the natural world, especially in its wildness, without discovering in the process corresponding regions within—after-images but also more than this. As Richard Sha and Joel Faflak argue in their illuminating volume *Romanticism and the Emotions* (2014), "emotion is the matrix through which the world is brought to our sensoria; it registers our response to this world; it worlds our world and thus makes sense of sense" (1). The emphasis here is on a subjective response through which the world is filtered, ordered, and then integrated into existing cognitive and emotional schema. But this is not to suggest that an encounter with what is new or hitherto uninterrogated or uncharted—an expression of wildness in physical space—invariably culminates in familiarization. The mind is not an organ of mere Procrustean reflexes; it also shifts its bearings, albeit often gradually, in response to stimuli that exceed our immediate protocols of integration. Romantic-era audiences, for example, owed a burgeoning lexical catalogue of elevated emotions—and perhaps a disposition to the emotions themselves—to the mapping of the sublime. Indeed, even before Edmund Burke's ground-breaking treatise on the subject, John Baillie's *An essay on the sublime* (1747) had already prepared eighteenth-century readers for an expansion of the inner world from encounters with natural immensity:

> every Person upon seeing a grand Object is affected with something which as it were extends his own Being, and expands it to a kind of immensity. Thus in viewing the Heavens, how is the Soul elevated; and stretching itself to larger Scenes and more extended Prospects, in a noble Enthusiasm of Grandeur quits the narrow Earth, darts from Planet to Planet, and takes in worlds at one View! (4)

Baillie's metaphors of emotional elasticity serve to develop a mental geography that is unconstrained by the laws of the physical universe and thus more replete with, or open to, areas of uncharted, even cosmic, wildness. A discrete sensory encounter—"seeing" a single "Grand object"—here sets in motion a process of expansion, a multidirectional dilation of experience, that leads the subject at once deeper within and further without. Burke, although he localizes the feeling of astonishment more directly in the overwhelmment of particular sensory organs—the eye, for example, which in response to grand objects, "vibrat[es] in all its parts . . . and consequently must produce an idea of the sublime" (132)—corroborates Baillie's account of an inner elevation that radically expands one's spatial and experiential horizons. Commenting on the inestimable value of "a search into the human mind" potentiated by encounters with sublime scenery, he traces the ultimate ascent of human enquiry directly to "the counsels of the Almighty by a consideration of his works" and concludes that "This elevation of the mind ought to be the principal end of all our studies" (35). In Baillie's "Heavens" Burke finds God and thus turns mere astonishment to worship.

Yet in the complex emotional matrix of Romantic-era writing, the outpourings of sublime enthusiasm are but one expression of what we might designate as extreme, unregulated, or "wild" feelings. Wordsworth's well-travelled delineation of "good poetry"—and the creative process more generally—as "the spontaneous overflow of powerful emotion" ("Preface" xiv) cannot be dismissed in this context as mere Romantic myth-making. Of particular interest in his formulation is the implied structure that precedes, one might even say enables, "overflow." Indeed, this latter word discloses in its prefix an assumption of unacknowledged forms or boundaries whose function is perhaps best understood as giving shape to and orienting, rather than prohibiting, the expression or trajectory of overabundance. To put it another way, wildness and cultivation whether in the realm of feelings or spaces, inner or outer geographies, have a mutually conditioning relationship.

That idea lies at the heart of one of the period's most important but also critically neglected reflections on poetry, emotion, and wildness: Francis Jeffrey's 1814 review of Byron's "The Corsair" and "The Bride of Abydos" in the *Edinburgh Review*. Six years before the publication of Thomas Love

Peacock's "The Four Ages of Poetry," Jeffrey undertakes a similarly controversial project in mapping a sequence of poetic development that includes the writers of his own day and age. Like Peacock, Jeffrey attunes his analysis to the regressive elements of contemporary poetics, yet his conclusions about their value differ sharply. Acknowledging that "the general history of poetry" unfolds in a "cycle," he distils its distinguishing variable as a slow progress of emotions from "violent" passions in "rude ages" to a "refinement of manners" which "exclude[s] the coarseness and offence of unrestrained and selfish emotions" (199). Over time, however, this refinement gives way to "the petty pretensions and joyless elegancies of fashion," out of which arises a renewed "avidity for strong emotions, which cannot be repressed" (200). The result, according to Jeffrey, is that "The feats of chivalry, and the loves of romance, are revived with more than their primitive wildness and ardour" (201). In Byron he finds the exemplar of this intensified "wildness": a poet who "delineate[s] with unequalled force and fidelity, the workings of those deep and powerful emotions which alternately enchant and agonize the minds of those who are exposed to their inroads" (198). An appetite for such emotions, for inner wilding, becomes for Jeffrey "the true characteristic of this age of the world" (201).

Jeffrey's conclusion, removed from its dynamic context of successive revolutions, has shaped an enduring narrative of the Romantic period, one that is rehearsed by Roderick Frazier Nash in *Wilderness and the American Mind* (1967) as well as Christopher Thacker in *The Wildness Pleases: The Origins of Romanticism* (1983)—arguably the two most influential twentieth-century reflections on the role of the wild in the Romantic imaginary. For both scholars, Romantic wildness, rooted in the aesthetic of the sublime but expressing itself in discourses social, cultural, and political, precipitated an investment in spaces and feelings remote from and critical of civilized life. Thacker's distillation of this retreat is particularly telling: "These writers, artists, *dilletanti* or amateurs in the eighteenth century were all trying, clearly or confusedly, with or without intelligence, and with degrees of sincerity and hypocrisy . . . to escape from what they were" (197–198). Thacker goes on to associate their ontological crisis with life in the city but eventually resolves upon the word "artificial" (198) to capture the essence of "what they were"—and presumably what wildness is not. Yet if we return to Jeffrey's cyclical history of poetry and to Wordsworth's valuation of emotional "*over*flow" in poetic expression, we can see that the "wild" is not simply a contradiction, indictment, or disavowal of the restraints of convention and cultivation, whether they emanate from urban life or a lived artificiality. The essays in this collection reveal a continual interplay between the wild and the ordered, the self-willed and the regulated, a crossing of boundaries—even a coinherence—suggestive of dynamic interrelation. As Van Horn suggests, "wildness . . . is an ongoing relationship, one in which human cultures—through active participation and humble restraint—become attuned to the community of life" (4). For writers of Anglo-European Romanticism, this relationship reveals varying levels of attunement and, in turn, shifting understandings of what constitutes, and how best to live in, a "community of life"—a challenge, we hasten to add, in which twenty-first-century readers are likewise vitally implicated.

Works cited

Baillie, John. *An Essay on the Sublime*. R. Dodsley, 1747.

"Bewilder." *Oxford English Dictionary*. Accessed 24 June 2020.

Bode, Christoph and Jacqueline Labbe, eds. *Romantic Localities: Europe Writes Place*. Routledge, 2016.

Burke, Edmund. *A Philosophical Enquiry into the Origin of our Ideas of the Sublime and Beautiful*. R. and J. Dodsley, 1757.

Butler, Tom. "Lives Not Our Own." *Keeping the Wild: Against the Domestication of Earth*, edited by George Wuerthner, Eileen Crist, and Tom Butler. Island P, 2014, pp. ix–xv.

Clark, Steve H. "Introduction." *Travel Writing and Empire: Postcolonial Theory in Transit*, edited by Steve H. Clark. Zed Books, 1999, 1–28.

Comet, Noah. *An Electronic Concordance to Keats' Poetry*. Romantic Circles, 2005, https://romantic-circles.org/reference/keatsconcordance/index.html.

Cronon, William. "Introduction: In Search of Nature." *Uncommon Ground: Rethinking the Human Place in Nature*, edited by William Cronon. W.W. Norton, 1996, pp. 23–68.

———. "The Trouble with Wilderness; or, Getting Back to the Wrong Nature." *Uncommon Ground: Rethinking the Human Place in Nature*, edited by William Cronon. W.W. Norton, 1996, pp. 69–90.

Hoad, T. F. "Bewilder." *The Concise Oxford Dictionary of English Etymology*. Oxford UP, 2003.

International Union for the Conservation of Nature (IUCN). "Category 1B: Wilderness Area," https://www.iucn.org/theme/protected-areas/about/protected-area-categories/category-ib-wilderness-area#:~:text=Protected%20areas%20that%20are%20usually, to%20preserve%20their%20natural%20condition.

Jeffrey, Francis. "Review: 'The Corsair' and 'The Bride of Abydos' By Lord Byron." *Edinburgh Review*, vol. 23, no. 1814, pp. 198–229.

Johnson, Samuel. *A Dictionary of the English Language*, 2 vols. W. Strahan, 1755.

Keats, John. *The Major Works*, edited by Frank Kermode. Oxford UP, 2008.

Kidner, David. "The Conceptual Assassination of Wilderness." *Keeping the Wild: Against the Domestication of Earth*. Ed. George Wuerthner, Eileen Crist and Tom Butler. Island P, 2014, pp. 10–15.

Leask, Nigel. "Romanticism and the Wider World: Poetry, Travel Writing and Empire." *The Cambridge History of English Romantic Literature*, edited by James Chandler. Cambridge UP, 2009, pp. 271–292.

Marris, Emma. *Rambunctious Garden: Saving Nature in a Post-Wild World*. Bloomsbury, 2013.

Mayhew, Robert. "The Character of English Geography c.1660–1800: A Textual Approach." *Journal of Historical Geography*, vol. 24, no. 4, 1998, pp. 385–412.

McKusick, James C. *Green Writing: Romanticism and Ecology*. Palgrave Macmillan, 2000.

———. "Introduction." *Romanticism and Ecology*. Romantic Circles, 2001, https://Romantic-circles.org/praxis/ecology/mckusick/mckusick_intro.

Morton, Timothy. *Ecology without Nature: Rethinking Environmental Aesthetics*. Harvard UP, 2007.

Nash, Roderick Frazier. "Wild World." *Keeping the Wild: Against the Domestication of Earth*, edited by George Wuerthner, Eileen Crist, and Tom Butler. Island P, 2014, pp. 183–187.

———. *Wilderness and the American Mind*. Yale UP, 1967.

Nichols, Ashton. *Beyond Romantic Ecocriticism: Toward Urbannatural Roosting*. Palgrave, 2011.

Purdy, Jedediah. *After Nature: A Politics for the Anthropocene*. Harvard UP, 2015.

Scott, John. *The Christian Life: From Its Beginning to Its Consummation in Glory*, Vol. 3. 5th ed. R. Ware et al., 1647.

Sha, Richard and Joel Faflak. "Introduction: Feeling Romanticism." *Romanticism and the Emotions*, edited by Richard Sha and Joel Faflak. Cambridge UP, 2014, pp. 1–18.

Smethurst, Paul. "Introduction." *Travel Writing, Form, and Empire: The Poetics and Politics of Mobility*, edited by Julia Kuehn and Paul Smethurst, Routledge, 2009, pp. 1–18.

Tegg, Thomas. *The London Encyclopedia: A Universal Dictionary of Science, Art, Literature, and Practical Mechanics. Comprising a Popular View of the Present State of Knowledge*. Tegg, 1829.

Thacker, Christopher. *The Wildness Pleases: The Origins of Romanticism*. St. Martin's, 1983.

Turner, Katherine. *British Travel Writers in Europe, 1750–1800: Authorship, Gender, and National Identity*. Ashgate, 2001.

United States Department of Justice. *The Wilderness Act of 1964*, https://www.justice.gov/enrd/wilderness-act-1964.

Van Horn, Gavin. "Introduction: Into the Wildness." *Wildness: Relations of People & Place*, edited by Gavin Van Horn and John Hausdoerffer. U of Chicago P, 2017, pp. 1–8.

Vannini, Phillip and April Vannini. *Wilderness*. Routledge, 2016.

Warren, Charles. "Wildness." *International Encyclopedia of Human Geography*, 2nd ed. Vol. 14. Elsevier, 2020, pp. 281–289.

Wiley, Michael. *Romantic Geography: Wordsworth and Anglo-European Spaces*. Palgrave, 1998.

Wordsworth, William. *A Guide through the District of the Lakes in the North of England*, edited by Nicholas Mason, Paul Westover, and Shannon Stimpson. Romantic Circles, 2020. https://Romantic-circles.org/editions/guide_lakes/editions.2020.guide_lakes.1835.html.

———. "Preface." *Lyrical Ballads, with Other Poems*. 2 vols., 2nd ed. Longman and Rees, 1800, pp. v–xlvi.

1 Weakness and wildness in Wordsworth's "The Brothers"

Emma Mason

Before the main dialogue of Wordsworth's "The Brothers," the Priest of Ennerdale describes Leonard Ewbank as "one of those who needs must leave the path / Of the world's business to go wild alone" (lines 102–103). Like the Priest's other conjectures about Leonard—that he is a stranger, a tourist, an idler—the designation "wild" hardly seems appropriate as the narrative unfolds. A mariner who has returned to his old village to reconnect with his brother, James, Leonard moves from a state of curious displacement to one of fraternal grief as the Priest relates the deaths of his father and brother. By the end of the poem he is the very opposite of a wild man, and assumes the demeanour of an ostensibly weak and tragic figure. Yet Leonard's weakness comes into focus as a virtue when read in relationship to his assumed wildness. Broken by his exile in the wilderness spaces of the sea and the colony, Leonard's separation from his family empties him of any strong or wild sense of self, will, or entitlement. This chapter reads this withdrawal from wildness through kenosis, a word that describes Christ's self-emptying of his divinity in Philippians 2:7. As a form of kenosis, Leonard's weakness can be read as a mode of relationship and affection that serves as an example to the Priest, whose dismissal of the stranger betrays his vocation and mission. As Leonard writes in a letter to the Priest sent following their encounter, it "was from the weakness of his heart, / He had not dared to tell him, who he was" (lines 427–429). These words have invited disdain from some critics. Charles Rzepka, for example, calls them a "half-hearted" and "vague and inadequate" rationale for Leonard's unwillingness to reveal his identity during his meeting with the Priest (81–82). But their fragility leaves an imprint of Leonard's weak and loving disposition on both the initially strong-willed Priest and the poem's narrative. For his weak approach to his village and the Priest presents the reader with a pastoral model of affection in which the wilderness moves its inhabitants into weakness and replaces certainty with dialogue. Wordsworth's pastoral thus goes beyond a defence of displaced small-holders and their affection for the land to embrace the experiences of all who live or visit rural sites, however contradictory (Wordsworth, *Letter* 322). The deliberately unresolved ending of the "The Brothers" (like its twin pastoral "Michael") transforms the desire for certainty into the acceptance of ambivalence as the basis for open exchange in which those involved become attuned to the emotional lives of each other.

This chapter begins by making an argument for weakness both as the consequence of time spent in the wilderness and as a methodological mode of enquiry based on Eve Sedgwick's weak reparative theory and Gianni Vattimo's weak thinking. Both promote the weak claims of charity and love: Sedgwick in relation to "strong" critics who, having wrestled their meaning from the text, are reluctant to admit the readings of others; and Vattimo in relation to the authoritative truth of God, which kenosis replaces with forgiveness, hospitality, and generosity. Sedgwick and Vattimo set the stage for my reading of Wordsworth's low and rustic pastoral form as "weak," one that leads readers, not into clarity and truths that restrict further thought and interpretation, but into uncertainty as a mode of possibility. As Fiona Stafford argues, Wordsworth's revival of pastoral contributed to a cultural promotion of life, hope, and health founded on plainness and simplicity. Restoring the form to "its original habitat, among real sheep and pasturage," she writes, Wordsworth strips it of what he called the "motley masquerade" of poetic "tricks" and "fancies" to reinstate a gentler language of kinship and sympathy (Wordsworth, "Appendix" 299; Stafford 121). But this language is also shaped by Christian humility, a trait rarely apparent in Wordsworth's self-identified religious characters like the Priest of Ennerdale, but unknowingly made available to them by resigned and patient figures like Leonard. As Wordsworth argued in *Essays on Epitaphs*, to which I turn in the second half of the chapter, the "strong" views of the assured and certain are more suited to cliché, satire, and pastiche, which often violently reduce the lives and philosophies of the weak to the level of caricature and so deem them redundant and expendable. As itself a weak mode, however, pastoral negates the strong structures of metaphysics to sustain weak moments of relationship and interpretation in which readers are invited to reflect and respond rather than review and decode. In this light, pastoral becomes more than a series of what Rob Nixon calls "redemptive silences" within a "spiritual geography" that overlooks the "historical hauntings" on which it is founded (238). Nixon's dismissal of its consolatory possibilities ignores the weak strategies the form offers poets to engender moments of affection between humans and nonhumans, strangers and friends. As I argue here, it is the uncertainty and loss that Leonard encounters in the landscape and wilderness of his rural home that softens him to what he finds there and enables his dialogue with the irritable Priest.

If the relationship between wildness and wilderness becomes a site for the exploration of ideas that help to define the Romantic period as this volume illustrates, then the religious and spiritual are central to it. As Laura Feldt argues, ideas of wilderness—forests, deserts, oceans, mountains—define the history of religion and belief as the faithful seek vital and raw spaces in which to experience the soul, God, the divine, and the spiritual self (1). Like Moses, both John the Baptist and Jesus walk into wilderness spaces to find redemption through God's presence following periods of trial, suffering, and defamiliarization. For David Jasper, the desert wilderness in particular offers an invitation to meet God by imitating and so participating in Jesus's suffering and meditation there (15). The desert, like the ocean, reminds the individual that he or she is powerless in the face of its vast and ancient immensity and must relinquish strong claims on the self. This process of self-forgetting or self-emptying is called kenosis in Christian tradition, a Greek term that describes the individual willingly throwing off aspirations to power and authority to embrace weakness and vulnerability. Paul uses the word kenosis to describe the incarnation, an event in which Christ temporarily "empties" himself of divinity to become Jesus the human and entirely receptive to God's will. Paul describes this transformation in his Epistle to the Philippians as a paradigm for his readers to follow, and encourages them to enter into kenosis through the imagination: "Let this mind be in you, which was also in Christ Jesus: Who, being in the form of God, thought it not robbery to be equal with God: But made himself of no reputation, and took upon him the form of a servant, and was made in the likeness of men" (King James, Phil. 2:5–7). For many readers of Philippians, kenosis reveals Christianity as a religion of humility, forgiveness, and patience through its instruction to renounce power, authority, and transcendence. Gianni Vattimo in particular turns to kenosis as the basis for his philosophy of *pensiero debole* or weak thinking, in which he promotes the weak claims of charity and love over the strong truth of God (*Belief*, "Weak Thought"). Like the theologian John Caputo, who also celebrates the weakness of God, Vattimo's philosophy makes uncertainty and negation the condition for thinking and reflection as well as kindness and sympathy (Caputo, *Weakness*).

In this context, weak thinking means interpreting without certainty in a manner analogous to Keats's negative capability wherein we dwell in "mysteries" and "doubts" and refrain from the reach for "fact and reason" (60). This idea has necessarily troubled some readers, for whom the turn to weakness translates into precarity and indecision, or worse, relativity and a lack of conviction. Yet for those who endorse weak thought and the modes of expression that emerge from it, weakness holds the potential to overcome dogmatic, binary, and fixed methodologies that are, as Sedgwick argues, both paranoid and suspicious (124–125). For Sedgwick, there is nothing wrong with paranoid reading until it becomes so dominant in critical approaches to texts that "to theorize out of anything *but* a paranoid critical stance" is dismissed as "naïve, pious, or complaisant" (126). Paranoid reading is in effect "strong" reading, and powerfully (and often defensively) overshadows and disavows other approaches to pose "as the very stuff of truth" (133, 135, 138). As such, strong reading and thinking are not only monopolistic, but symbolically violent, intent as they are on the exposure (and then erasure) of that with which they disagree as irrelevant and obscure (141). Sedgwick offers reparative reading as a weak counter, a queer cultural mode which privileges mistakes, pleasure, and "soft" reform as the basis for a critical practice (and explicitly not a theoretical ideology). While she in the end favours an approach that brings the paranoid and reparative together to "interdigitate," the hermeneutic foundation of such relationship is necessarily weak because any intervention by strong thought neutralizes its opposite. Thinkers such as Paul Ricœur, Peter Elbow, and Kathleen Fitzpatrick concur, explicating suspicious, cynical, and doubtful approaches to reading and interpretation in order to defend generosity, subjectivity, creativity, and belief. As Fitzpatrick argues in her defence of "generous thinking," only those who practise a nondualist and charitable mode of criticism can bring critical distance and emotional response together (107). Strong thinkers, by contrast, dismiss the affective to "wrest" meaning from texts and then furiously and competitively defend their discoveries. As Stephen Best and Sharon Marcus argue, this forceful approach, exemplified by critics like Harold Bloom or Frederic Jameson, regards contradictions and tensions as "clues" to otherwise veiled historical and political truths, and not precursors to uncertain dialogue (7).

Even if critics could agree that criticism was a "strenuous and heroic" form of labour, the certainty this perspective requires is arguably no longer available. Timothy Morton's term "hyperobjects" underlines the redundancy of strong, assured thought based on individual distinction. Comprising things and phenomena that massively transcend particular spatiotemporal contexts, hyperobjects describe anything that transforms and eventually outlasts the lives of human beings—global warming, polystyrene, radiation, the biosphere, plutonium, oil reserves, viruses. They thus reveal human sovereignty and claims to power as a fiction, one that begins to be dismantled in the late eighteenth century with the advent of steam. Morton's argument that viscous and nonlocal phenomena like steam and later oil simply exceed human attempts to control their influence on the world speaks to Wordsworth's willing relinquishment of human power for relational weakness. Romantic pastoral is a response to steam not because it nostalgically idealizes the land as somehow separate from the influence of hyperobjects, but because it accepts existence as an entanglement of different things, beings, and perspectives. Wordsworth's defence of weakness, then, resonates through Morton's assertion that humans should relinquish power to confront our own fragility and finitude and become more "mindful." By embracing a state of "awareness" and "simple letting-be," he writes, we will experience a "deep acceptance of co-existence" and "inner depth" through which to "comprehend infinity" and abandon "the search for ultimate men and supermen" (*Hyperobjects* 176, 199). The religiosity of the terminology used here—mindfulness, co-existence, infinity—is made more apparent when Morton concludes his book with a reference to Heidegger's infamous statement, "only a god can save us now": "We just don't know what sort of god," Morton writes (201). But his own discussion in which strong thinking and ultimate supermen are questioned by a willingness to sit with uncertainty suggests that he does know what kind of god might make sense of an undetermined world—a weak one. Vattimo directly answers his question by rephrasing Heidegger's statement as "only a 'kenotic' god can save us" (*Farewell* 47). For Vattimo, despite the best efforts of institutional Christianity to distort the weakness of its own belief system into a "strong" dogmatism, the death and kenosis of Jesus suggest otherwise (*Farewell* 63). Weak Christianity is not a metaphysics or set of sacred or dogmatic claims, then, but rather an ecumenical hermeneutic that infinitely changes our interpretive direction towards charity and love (*Farewell* 126).

Removed from the violent objectivism of metaphysics and the literalism of orthodoxy, Vattimo's Christianity is a charitable, desacralized approach to interpretation and relationship that welcomes conflicting and multiple perspectives (*After the Death* 99, 102). But he goes further to argue that only the poem can best articulate such diversity, a form that he suggests refuses the self-evident without collapsing into an interpretive jumble in which meaningfulness is excised. In his essay, "The Shattering of the Poetic Word," Vattimo argues that the undecidability of the poem allows the reader to become acquainted with and experience truths without rushing to affirm them. Poetry's resistance to fixed or self-evident meaning leads to a fracturing of the word that reveals language not as an instrumental means for "showing things," but as a way to experience being and mortality. Poetic language "opens up" our worlds, their possible horizons, and the "unfolded meanings" within them to then break them up (*End of Modernity* 71–72). While readers always engage with a poem in a particular moment, the poem's passage through time allows for infinite reinterpretations (including those that dismiss it as a text worth reading at all) and so creates the space for a weak thinking in which readers move closer to or further away from various meanings, ideas, and definitions. The poem thus serves as a "monument" or tombstone for Vattimo, that which inscribes meanings but leaves them to fade and crumble: the poet's effort to shape and "sculpt" the poem is thus "an anticipation of the essentializing erosion that time exerts upon the work" (*End of Modernity* 75). For readers of Wordsworth, Vattimo's description instantly recalls epitaphic moments in his poetry (the last two lines of "The Brothers," for example, to which I return below), and the opening statement to the first of his *Essays Upon Epitaphs*, in which Wordsworth writers that "an Epitaph presupposes a Monument" to unite "the two worlds of the living and the dead" (*Essays* 49, 60). Prominently located in the parish-church, "the visible centre of a community of the living and the dead," epitaphs bring together those who view them in a moment of quiet remembrance experienced "in the light of love" (*Essays* 56, 63). Like Vattimo's poem, the epitaph is irreducible to its ostensible meaning (its content suggests that there are no "*bad* People buried" in graveyards), but it is also kenotic and weak in that it encourages remembrance while emptying itself of assertion or judgement (*Essays* 63). As the precondition for weakness, the parish graveyard is itself a kind of wilderness in which certainty and order are exchanged for interaction and sympathy.

The most affecting epitaph Wordsworth discusses in his *Essays* is recognizably weak, engraved as it is on "a very Small Stone laid upon the ground, bearing nothing more than the name of the Deceased with the date of birth and death, importing that it was an Infant which had been born one day and died the following" (*Essays* 93). As an ontological statement about existence, these dates free the reader into an experience of the reality of this death—"awful thoughts of rights conferred, of hopes awakened, of remembrances stealing away or vanishing"—and so illuminate the depth of grief the living must negotiate (*Essays* 93). At the same time, this truth is one that might shift (or shatter) the more it is remembered and reflected on, and so transform into other meanings, such as gratitude for hopes and lives that are fulfilled or the solace of a supportive community. The dates thus do exactly what Wordsworth requires of language: they incarnate rather than "clothe" thought by moving the reader away from artificial, mannered, and laboured epitaphs to a textual presence that requires stillness and silence. This much cited passage from the third essay, in which language, "if it do not uphold, and feed, and leave in quiet" turns into a violent and wild "counter-spirit," "noiselessly at work to derange, to subvert, to lay waste, to vitiate, and to dissolve," is often read through suspicion, if not paranoia, as contradictory and conflicted (*Essays* 85). Yet read through kenosis, the statement that language should work to sustain meanings that quietly nourish the reader speaks to a moral responsibility to communicate without destroying the perspective of those with whom one disagrees. Vattimo almost restates Wordsworth's commentary in his suggestion that communication free of "metaphysical rigidity" conceives meaning (and so ethics and morality) as weak "negotiation" (*Transparent* 118). It is this weakness that I argue Leonard embodies and that the poem more broadly promotes through its dependence on weak forms of communication and response. The second part of this chapter explicates Wordsworth's defence of Leonard's weakness, one learned through his time in the wilderness, and realized in the poem's internal critique of metaphysical and orthodox rigidity as figured through the Priest.

From the opening lines of "The Brothers," the Priest of Ennerdale is presented as a cheerful but disparaging figure who immediately condemns Leonard as a lounging sightseer. The Priest's frustration is palpable: he is averse to anyone he does not immediately recognize as part of his parish, and is particularly resistant to "idle" tourists for whom his churchyard promises aesthetic pleasure. That he ought to recognize his former parishioner is an uncomfortable aspect of the narrative that resounds throughout the ensuing dialogue between the two men. As the Priest remains in ignorance, the reader learns that Leonard once lived in Ennerdale with his family, to whom he wishes to return following "twenty seasons" on the sea (line 41). Set against the increasingly impatient Priest, now eager to "accost / The Stranger," the character of Leonard is a gentler and more visionary presence, "half a Shepherd on the stormy seas" for whom the "broad green wave and sparkling foam" would flash up images of the former mountains, hills, and sheep with whom he spent his childhood (lines 35–36, 43, 53). Following Alan Bewell, readers regard this passage as a description of calenture, and Leonard "in danger of falling victim" to a nostalgic "fiction" in which he drowns "in a landscape of his own making" (60). While his colonial encounter with the "Indian Isles" drives his longing for an idealized pastoral and domestic home, Leonard does not sustain this fiction on return to Ennerdale. Rather, he is weakened on arrival in the village: the narrator conveys that "his heart / Fail'd in him" as he approaches his former home and he remains confused and disoriented by the nameless graveyard of which the Priest is so proud (lines 74–75). This confusion is not presented negatively, however. His memory of losing his way earlier that afternoon comforts Leonard, who is increasingly at ease with the uncertainty of his present moment, a stark contrast with both the instrumental aims of his abandoned commercialism on the sea and the assured convictions of the Priest. The Priest's mannerisms, language, and body language in the first half of the poem in particular paint him as a censorious figure who abruptly "stop[s] short" at the "church-yard gate" to view Leonard, and smiles complacently before executing a startling misreading of him (lines 98–99). He assumes the stranger is a self-involved hedonist, "one of those who needs must leave the path / Of the world's business to go wild alone" and eager to follow "his fancies by the hour" (lines 102–103, 106). This undiscerning reading is even more misplaced when compared to Leonard's uncertain but careful perception of their interaction, which the narrator presents as a far more reliable compass to its affective content.

Leonard's affective intuition is further apparent in one of the poem's pivotal narrative devices—that the Priest fails to recognize him even as he remembers the Priest. Leonard's reading of his former vicar as "quiet" and "peaceful" is perhaps generous in light of the latter's curt welcome, but it is consistent with his propensity to find the weak and gentle in others, a forebear of Fitzpatrick's generous thinking (lines 118–119). It also introduces a recurrent motif in the poem, namely, weak moments of relationship and interpretation that serve as affective touchstones to map the story for the reader. These moments move us towards an understanding of the poem in which incomplete and unresolved details offer a way out of conclusive judgements rather than pointing to stylistic or narrative failure. In other words, "The Brothers" resists diagnostic exposition to free itself from what Vattimo calls the categorizing, measuring, and framing impulses of strong structures like the Enlightenment that equate weakness with deficiency ("Foreword" xiv). The dialogue between the Priest and Leonard, for example, stages an interaction between the former's desire to "chronicle" the history of Ennerdale and the latter's willingness to listen incognito (line 160). On the surface, this appears like a struggle between two forms of weakness—the Priest uninterested in "facts or dates" to relate the history of the village, and Leonard hiding his identity in order to indirectly learn about his family (line 159).

Yet the Priest's refusal to rely on "symbols" or "names and epitaphs" in his churchyard is a strong, categorizing move (lines 176, 179). The village ostensibly has "no need" of gravestones because its members talk with each other about the dead (line 176). But the absence of any symbols or epitaphs necessarily excludes anyone not immediately involved in their community—like Leonard—as well as anyone who might wish to mourn the dead in the dark. Only through Leonard's sensitive and attentive engagement with the Priest can he conceptualize the "second life" the dead possess in the memories of his former community (line 183). Self-emptied of ego and so receptive to those he encounters, Leonard enacts exactly what the Priest ought to represent but does not: openness and hospitality. While Leonard is not in fact a "stranger on this land" (*King James*, Ps. 119), he accepts this status in order to embody the powerlessness and vulnerability learned through his former experience in the wilderness.

The link between wilderness and weakness is further deepened by the Priest's description of Leonard's father Walter and brother James. Walter and James are both depicted as suffering their own wilderness experiences, Walter through his harsh struggle "with bond, / Interest, and mortgages," and James in relation to his sleepwalking (lines 212–213). Like Leonard, his father and brother are opened into sustained weakness through their time in the wilderness. The Priest appears to almost comprehend Walter's weakened being by splitting him into two personas or "Two fathers in one father"—the labourer and the caring parent. But he has to feminize Walter to do so, describing him as "half a mother" and thus gendering weakness as female (lines 228, 233). Leonard, however, is more comfortable with his weakened status: he weeps and is visibly overcome by the Priest's narration, which becomes progressively more sympathetic as he witnesses Leonard's response. The young Leonard is already saint-like in the Priest's account, in which he appears tenderly caring for James and, during their ambles in the mountains, "Bearing his Brother on his back" in an echo of St. Christopher carrying the child Christ (line 255). Indeed, the two brothers are like beloveds to each other, twin "springs which bubbled side by side" and "Roebucks" that bound "o'er the hills" (lines 138, 273) in a direct reference to the Song of Solomon's description of lovers "leaping upon the mountains, skipping upon the hills. My beloved is like a roe or a young hart" (*King James*, Song of Sol. 2:8–9). The Priest frames these references with further religious comments—God as the maker of the book of nature, for example, and the Priest's gift of a Bible to Leonard before he left Ennerdale—both of which Leonard notably interrupts. For he wishes to push past the Priest's nostalgic tale of a parish of able workers and good church-goers into stories about relationship, emotion, and familial love. After all, his family's former obsession with work as a moral good is what justifies Leonard's departure from his community in the first place, and specifically his abandonment of James. Following the death of his father, who is left "too weak" (with the emphasis on "too" here) after his investment in the estate and their home,

Leonard takes up the offer of work from an uncle "chiefly for his Brother's sake" (lines 295, 301). His silence on his life away from Ennerdale is an implicit critique of the horrors of trafficking and slavery: as a wilderness experience, it is one that breaks rather than simply weakens those within it. Leonard's redemption comes through his love for James, the brother from whom he learned weakness as a child, and whose memory returns him to weakness from brokenness even after his death.

James's kenosis is, like Christ's, one that culminates in death, and, if not resurrection, an ascension through his continued weak influence on the village. A spirited "Mountain Boy" and "Child of all the dale," James is directly associated with wilderness (lines 332, 339). His decline following Leonard's departure provokes an outpouring of affection from the community, who shelter, feed, and take care of him. But his pining for Leonard culminates in a serious case of somnambulism, in which he habitually sleepwalks in search of his beloved sibling. Mourning the loss of Leonard, whom he and the village assume is now dead, James seeks solace in the mountains and one day sleepwalks off the summit known as "The Pillar." While the Priest is himself weakened at this point in the narrative by the visibly moved Leonard, he is unable to resist embellishing his story with the grisly details of James's "mangled limbs" and details of his burial the "third day after" he was found (lines 378–379). But James's weakness is redeemed here: while he is buried rather than rises again on the third day like the crucified Christ, the echo of the resurrection tells the reader that James lives again in the affection the villagers have for him. This is in part because, unlike the dead buried in the sign-less churchyard, James's death is marked by the also Christ-like symbol of his "Shepherd's staff," which "caught" on the side of the rock from which he fell, and hung "there for many years" before it "mouldered" away (lines 400–403). While this is not a permanent tombstone, it is a shattered one that has since materially vanished, but is spiritually still apparent to those who look on "The Pillar" and remember James. The staff also brings the Priest's narrative to a close and both he and Leonard fall into silence. The Priest's rambling, occasionally insensitive, and sometimes elaborated story stands in stark contrast to Leonard's minimal and slight utterances: where the Priest "clothes" his thoughts with too many words, Leonard invariably "incarnates" his emotions and grief. If the Priest's narrative does not quite derange or dissolve the affective meaning of his account, neither does it nourish as do Leonard's words. One might even imagine a kind of sub-poem comprised only of Leonard's contributions, a text able to sustain the affections on which "The Brothers" is focussed, albeit weakly without the intrusion of the Priest's "strong" narration.

While the Priest is gentled by Leonard's visible grief at James's death, it is not enough for him to recognize his former parishioner. Leonard, however, is almost paralysed by grief at this point in the poem, and both men depart without further conversation. Yet like Wordsworth's epitaph that bears only the name of the deceased with the date of birth and death, Leonard's concluding and scarce sentence, "My Brother," discloses more than the Priest's entire recital:

> The Stranger would have thank'd him, but he felt
> Tears rushing in; both left the spot in silence,
> And Leonard, when they reach'd the church-yard gate,
> As the Priest lifted up the latch, turn'd round,
> And, looking at the grave, he said, "My Brother."
> The Vicar did not hear the words. (lines 403–408)

The Priest's failure to hear Leonard corresponds to his broader inability to see him. There is no question that the Priest warms to him through their interaction, and even invites Leonard in for supper. But he effectively breaks the bonds of the community his narration eulogizes because of what Paul Magnuson calls his "simple literalism," one that leaves Leonard a stranger "separated from local and public knowledge" (260). The "festival" the Priest promises would joyously greet the returned Leonard is traded for misrecognition and then silence, after which the reader is left only with Leonard's reflections on his brother and home (line 308). The tragedy of the poem is pronounced here, and yet there is another way of reading its conclusion that recognizes redemption, if not for Leonard, then at least for the Priest. Leonard's thoughtful, kind, and attentive conduct not only reverses the Priest's initial perception of him, but reminds him of the duty his vocation requires: to be hospitable towards strangers. That this moment of realization comes as a result of Leonard's weakness is emphasized in the poem's final lines, in which the spotlight falls on Leonard alone. Still reeling from the news of James's death, Leonard stops by a grove and pauses under the trees to review "All that the Priest had said":

> his early years
> Were with him in his heart: his cherish'd hopes,
> And thoughts which had been his an hour before,
> All press'd on him with such a weight, that now,
> This vale, where he had been so happy, seem'd
> A place in which he could not bear to live:
> So he relinquish'd all his purposes.
> He travell'd on to Egremont: and thence,
> That night, he wrote a letter to the Priest,
> Reminding him of what had pass'd between them;
> And adding, with a hope to be forgiven,
> That it was from the weakness of his heart
> He had not dared to tell him who he was.
>
> This done, he went on shipboard, and is now
> A Seaman, a grey-headed Mariner. (lines 417–431)

Leonard has dashed his own "cherished hopes" here: unable to call himself a Ewbank in front of the Priest, he is dissociated from the vale and the community that might have offered him solace (line 418). Yet Leonard does reveal his identity in the end, and writes to the Priest that "it was from the weakness of his heart" that he remained anonymous during their dialogue (line 428). As Kurt Fosso argues, Leonard is reconnected to his former parish by becoming "a part of the community's exclusive economy of oral epitaphs" in the form of his encounter with, and then letter to, the Priest (153). For Fosso, this suggests that those who elegize the dead legislate for a community integrated by the "mourning-work" of the living for the lost (154). But the Priest is not fit for this role before his encounter with Leonard, and it is the latter's heartfelt weakness that transforms him from oral historian to compassionate pastor.

If "The Brothers" ended here it might fall prey to Nixon's critique of the pastoral, and leave the reader only with the "redemptive silence" with which the narrative closes. But the final two lines block any such reading. Many critics have noted the epitaphic nature of these lines, detached on the page and sharply divergent in tone and feel from the preceding emotional dialogue (Fosso 153; Wolfson 84). They also mark Leonard's return to the wilderness from which he arrived and suspend him there indefinitely: he does not simply go "on shipboard," but remains there until he is an old man, "a grey-headed Mariner" (lines 430–431). One might read these lines as a kind of obituary for Leonard, who is now superfluous to the narrative following his renewal of the Priest's vocation. They also intimate Leonard's retreat from the weakness he incarnated in the dialogue and letter alike, and consequent surrender to the certainty that life as a mariner guarantees. The sea is no longer a wilderness from which the individual emerges emptied of self and ambition, but a place of inevitable employment. The poem appears to close with this acquiescence to a strong existence defined by the colonialism and trade from which Leonard initially fled. Any possibility that he might reappear in Ennerdale is completely voided by this pragmatic curtain fall, one that parallels Michael's paralysis as he sits with his dog at the unfinished Sheep-fold at Green-head Gill. Yet, like Michael, who offers the reader hope through his love for his dog and family, Leonard leaves behind a reparative and mindful quietude that offers more than spiritual strength to the Priest—it reminds him for whom the churchyard and Parish Chapel exist: the strange, weak, and dispossessed. Leonard's departure also affirms how difficult it is to sustain a weak outlook and disposition in the world, one that even James only preserves because of his early death.

Weakness is presented as a virtue in the poem, then, but one that needs to "interdigitate," to use Sedgwick's term, with strength and wildness, rather than swing fully back into an egoic intransigence. It is perhaps this intermediate state that the Priest comes to enjoy, one in which he is silent and reverential before Leonard's grief, but gracious and accommodating as he invites him in for supper. His initially strong reading of the stranger is now fully modulated by Leonard's weakness, which simultaneously serves as a reminder that the religion he serves is a weak one in which the only law is gift and charity (Caputo 235). At last freed from his own imprisonment in a wilderness of judgement and admonition, the Priest embodies a generative strength founded on humility and *agape* and "made perfect in weakness" (2 Cor. 12:9). While Leonard might not be able to sustain such balance, the poem suggests that the newly buoyed Priest can, devoted as he is to a God emptied of will and determination. The reader too is immersed in the wilderness space of the poem, one in which she is moved into a weak and charitable encounter with the text remote from the "motley masquerade," not just of linguistic "tricks," but also of strong and paranoid criticism. Wordsworth's pastoral might not offer direct spiritual guidance to the reader, but weakness is offered as a possible salve for what he called the "decay of the domestic affections" in his famous letter to Charles James Fox (306). Only by acknowledging the ways in which others "differ from us," he writes, will those affections be restored (308). "The Brothers" mediates that difference by celebrating forms of relationship and interpretation based on weakness and affection, and which obviate the power of the strong, if not paranoid, thinker.

Works cited

Best, Stephen and Sharon Marcus. "Surface Reading: An Introduction." *Representations*, vol. 108, no. 1, 2009, pp. 1–21.

Caputo, John. *The Weakness of God: A Theology of the Event*. Indiana UP, 2006.

———. With Gianni Vattimo. *After the Death of God*. Edited by Jeffrey W. Robbins. Columbia UP, 2007.

Elbow, Peter. *Writing Without Teachers*. Oxford UP, 1973.

Feldt, Laura. "Wilderness in Mythology and Religion." *Wilderness in Mythology and Religion: Approaching Religious Spatialities, Cosmologies, and Ideas of Wild Nature*. edited by Laura Feldt, De Gruyter, 2012, pp. 1–23.

Fitzpatrick, Kathleen. *Generous Thinking: A Radical Approach to Saving the University*. Johns Hopkins UP, 2019.

Fosso, Kurt. *Buried Communities: Wordsworth and the Bonds of Mourning*. State U New York P, 2004.

Jasper, David. *The Sacred Desert: Religion, Literature, Art, and Culture*. Blackwell, 2004.

Keats, John. "Letter to George and Thomas Keats, 22 December 1817." *Selected Letters of John Keats*. Edited by Grant F. Scott. Harvard UP, 2005, pp. 59–61.

Magnuson, Paul. *Coleridge and Wordsworth: A Lyrical Dialogue*. Princeton UP, 2014.

Morton, Timothy. *Hyperobjects: Philosophy and Ecology after the End of the World*. U of Minnesota P, 2013.

Nixon, Rob. *Slow Violence and the Environmentalism of the Poor*. Harvard UP, 2013.

Ricœur, Paul. *Freud and Philosophy: An Essay on Interpretation*. Yale UP, 1970.

Rzepka, Charles. "Wordsworth and Lyrical Archaeology: The Poetics of Pre-Historical Imagination in 'The Brothers.'" *The Wordsworth Circle*, vol. 34, no. 2, 2003, pp. 81–85.

Sedgwick, Eve Kosofsky. *Touching Feeling: Affect, Pedagogy, Performativity.* Duke UP, 2003.

Stafford, Fiona. "Plain Living and Ungarnish'd Stories: Wordsworth and the Survival of Pastoral." *The Review of English Studies*, vol. 59, no. 238, 2008, pp. 118–133.

Vattimo, Gianni. *Belief.* 1996. Trans. Luca D'Isanto and David Webb. Stanford UP, 1999.

———. *The End of Modernity: Nihilism and Hermeneutics in Post-modern Culture.* Trans. Jon R. Snyder. Polity P, 1988.

———. *A Farewell to Truth.* Translated by. William McCuaig. Columbia UP, 2011.

———. *The Transparent Society.* Translated by David Webb. Polity P, 1992.

———. With Santiago Zabala. "Foreword." Michael Marder, *Plant Thinking: A Philosophy of Vegetal Life.* Columbia UP, 2013, pp. xi–xv.

Vattimo, Gianni, Santiago Zabala, and Yaakov Mascetti. "'Weak Thought' and the Reduction of Violence: A Dialogue with Gianni Vattimo." *Common Knowledge*, vol. 8, no. 3, 2002, pp. 452–463.

Wolfson, Susan. *The Questioning Presence: Wordsworth, Keats, and the Interrogative Mode in Romantic Poetry.* Cornell UP, 1986.

Wordsworth, William. "Appendix. 'What Is Usually Called Poetic Diction.'" *Lyrical Ballads 1798 and 1802.* Edited by Fiona Stafford. Oxford UP, 2013, pp. 298–301.

———. "The Brothers." 1800. *Lyrical Ballads 1798 and 1802*, edited by Fiona Stafford. Oxford UP, 2013, pp. 209–221.

———. "Letter to Charles James Fox, 14 January 1801." *Lyrical Ballads 1798 and 1802.* Edited by Fiona Stafford. Oxford UP, 2013, pp. 306–309.

———. "Letter to Thomas Poole, April 1801." *The Letters of William and Dorothy Wordsworth: The Early Years, 1787–1805.* Edited by Ernest de Selincourt, revised by Chester L. Shaver. Clarendon P, 1967, pp. 322–323.

2 Wild freedom and careful wandering in the poetry of William Wordsworth and John Clare

Sue Edney

Octavius Gilchrist, writing in *The London Magazine* in 1820, was keen to detail certain books that the "peasant poet" John Clare had in his possession: the Bible, a "single volume" of Pope, and Robert Bloomfield's *Wild Flowers, or, Pastoral and Local Poetry* (1806), a title that sums up the dilemma of nineteenth-century rural working-class poets (9). Intensely domestic, they were expected to be the voices of order and contentment, paradoxically epitomized in wild nonhuman nature. However, Gilchrist's choice from Clare's numerous books also creates an image that fits a middle-class audience's appreciation of the humble poet: god-fearing, yet a sensitive man of taste. Although he called himself a "bard of the wild flowers" towards the end of his life (Mahood 112–113), Clare was a comprehensive advocate for all wild nonhuman beings, plant or animal. Wild flowers, though, were especially loved by the reading public; they had no teeth or claws; they were not scaly, slimy, or wet; they were, in general, small. So, Clare's early sonnet "The Primrose," in *Poems Descriptive of Rural Life and Scenery* (1820), "was very much to the taste of the Town," observes Molly Mahood and helped to secure Clare's "public acclaim" (115): "How much thy presence beautifies the ground: / How sweet thy modest, unaffected pride" (lines 5–6). There is little of the wild in this poem, and much of the humble, tasteful plebeian who knows his place. And Clare certainly knew his place: loved it, resisted it, grieved for it, tried to escape it, and longed to return.

Clare's wild places are created out of cultivation; nearly all Clare's poems are about domestic spaces: his landscape was worked, not wild. Any wildernesses were on the edges, in the small woods, streams, heaths, and fens of his surroundings. These gave the appearance of being everyone's property or no one's; the majority of Clare's landscapes belonged to farmers and estate owners. William Marshall, writing in 1785, had already rejected the notion of a "natural" English landscape:

> Wherever cultivation has set its foot, —wherever the plow and spade have laid fallow the soil, —Nature is become extinct; and it is in neglected or less cultivated places, in mosses and mountains, in forests and parochial wastes, we are to seek for any thing *near* a state of Nature. (585)

Wildness was attached to unfettered, unbounded spaces, as well as to human and nonhuman others. Even as it might provoke anxiety, wildness as a concept also stirred envious desires for a life without constraint—restrictions imposed by convention and middle-class propriety for a man in William Wordsworth's position, or oppressive labour and extreme poverty in Clare's situation. Neither "wild" nor "free" carries the same affective signification for these poets, even when applied to the natural world which they both loved and celebrated; their different social and domestic situations necessitated differing interpretations of what might be acceptably wild and free. "Pastoral" implies safety, and although it is yoked to "wild" in Bloomfield's title, they are only flowers. However, the concept of freedom in its relationship to the wild connects Wordsworth's and Clare's philosophy of place-creation and entanglement in the "life of things," as Wordsworth epitomizes life-value in "Lines written a few miles above Tintern Abbey." In Wordsworth's meditation on human freedom in connection with the freedom of nature, the poet connects wild with free in language of the spirit as much as of nature. In particular, Wordsworth implies that the natural world can permeate a mind open to its power in such a way that "with an eye made quiet by the power / Of harmony and the deep power of joy, / We see into the life of things" (lines 47–49).

Things have stories to tell, matter is "a site of narrativity . . . where the world reveals its creative becoming, its dynamism and its reenchantment" (Oppermann 31). Wordsworth and Clare use different story-telling methods, yet their similarities in the context of "response-ability" are greater than their differences (Barad, "Interview" 55). Both poets, in other words, respond to matter in the form of human/nonhuman interactions in contexts that value "creative becoming" of things themselves and of their own poetic developments. Their narrative strategies, though, stem from different positions. Clare celebrates, and often incorporates in his writing, the oral wealth of stories from his parents, his neighbours and friends, alongside tales read in the "chap books" he loved as a child and which were often recreated in the eighteenth-century poems with which he grew up. Wordsworth deliberately places himself at one remove, looking for a means to unite middle-class literary experience with rural reality, in a genuine if at times clumsy attempt to elevate rustic values of frugality, simplicity, and natural beauty beyond the artifices of literary establishments.

Like many working-class writers such as Stephen Duck and Mary Collier, Clare saw benefits in translating his rural experience through Augustan poetic practices—using heroic couplets, for example, in settings that, while not celebrating England's imperial wealth, explored England's village lives at work and leisure. Simon Kövesi argues that a primary view of Clare as a poet of agricultural Helpston has the potential to be more limiting than Clare's actual restriction in Helpston or Northborough. "The awkwardness of Clare's landless occupancy of his place," writes Kövesi, "is sourced in an awareness that others in the poetic world," such as Wordsworth and

Byron whose work Clare knew and admired, "exhibit some choice about the spaces they move through" (39). Clare's restriction must be seen as a virtue however, as Kövesi argues, because it enables Clare to develop imaginative tactics to "write beyond his situation" (40) while simultaneously creating a public image of everyday rural life. As John Goodridge describes Clare's poetic choices, the "question of how to use eighteenth-century models to write about the hardship of the rural world, and the destruction wrought by intensification and enclosure, was a particularly difficult one. Clare drew on Goldsmith, Crabbe, Cowper, and the sensibility poets, as well as Bloomfield and Wordsworth, to do so" (270). Bloomfield was Clare's poetic hero, close to him in observational understanding of what it was to be a field-worker. When Giles rests in the field margins in Bloomfield's *The Farmer's Boy* (1800), he watches the "swarming insects creep around his head" ("Summer," line 74) with the same acute descriptive accuracy and personal involvement that characterize Clare's detailed accounts. For Bloomfield, there is a significant and suggestive freedom exhibited by a "dust-colour'd beetle" that "climbs with pain" among the long grasses ("Summer," line 75). Goodridge comments on the references Clare makes to other working-class poets in his own work: "There is a genuine sense of solidarity between these labouring-class men. They share intellectual and literary interests," and according to Goodridge, they also "share a common struggle to escape" from the confines of a class-based judgement of their literary merits.

However, the popularity of self-taught "peasant" and "artisan" poets in the late eighteenth and early nineteenth centuries presented a dilemma for Wordsworth; "[p]easant poetry" made his position "almost impossibly awkward," explains McEathron, "as he attempted, in the years around 1800, to establish himself as a poet. He was, after all, a writer of rustic verse who could not profess to be an 'authentic' rustic" (5). Wordsworth was as constrained by established aesthetic models as Clare, although from inside the establishment, and used some convoluted techniques in order to free his writing—and, in his opinion, the position of poetry as a transmitter of truth—from a different kind of ghettoization. Unlike Clare's "dispatches from the front," as John Ashbery characterizes many of his poems (17), Wordsworth's determined attempts to have the rural poor "speak" often render them oddly silent. His rural inhabitants are often isolated individuals, beggars, vagrants, ex-soldiers, and sailors, who wander through necessity as much as choice, and are met with on the road by a man who could afford to walk where he liked. Wordsworth "rarely represents labouring people engaging in the incidental chatter" of Clare's poetry, notes Simon White (152), or how "in some pleasant nook" the haymaking "swain and maid / Lean oer their rakes and loiter in the shade . . . Hark at that happy shout – and song between" ("Haymaking" 7–8, 11). In contrast to Clare, Scott McEathron argues that Wordsworth's "speakers cannot perform the task at hand," that of story-telling, without middle-class poetic intervention (16). Wordsworth's desire to raise poetic consciousness on behalf of rural themes leads him to create a linguistic relationship with wide-ranging materialities, resulting in an animating "language of things," as Adam Potkay argues (399), intended to be inclusive of the material and immaterial, earth and spirit. His "things" are "uncontainable by any narrow definition . . . [T]hey bespeak the fusion of object and event, matter and energy" (Potkay 391). By these means, even the simplest forms and subjects can be elevated to poetry of both moral and aesthetic value.

One of Wordsworth's difficulties in his poetic practice is that while the transcendent lyricism of his personal investigations into spirit and nature is covered by this ethical "language of things," much of his writing is narrative, and here Wordsworth stumbles into wild things with which Clare has quite a different relationship. However profound Wordsworth might profess his encounter with nonhuman things to be, it is hierarchical; human-related "things"—ethical and spiritual qualities and practices—are influenced by "natural" or "wild" freedoms and virtues for the ultimate benefit of the human. It is at some remove from Clare's rootedness in matter itself—his compulsive listing that created, for his critics, an over-descriptive poetry; his collecting of anything he found interesting; "to capture in relation to a specific environment or locale, whatever natural feature . . . he might witness," comments Kövesi. He adds that these "lists" might operate as "communitarian nodes—markers of inclusive connection" (203–204). In these aspects, Clare has more in common with Dorothy than William Wordsworth, who recreated in her journals the actualities of daily life in Grasmere in ways that her brother attempted to rise above.

Erica McAlpine considers Clare's intensely observational writing to be unusually objective in comparison to that of his Romantic contemporaries, who looked for "deep and spiritual knowledge of the world around them" (79). However, like Bloomfield's alter ego Giles, Clare deliberately immersed his poetry in the materiality of grass, insects, blossom, birds, and animals in order to distance himself from a perceived lack of deep and spiritual knowledge in the everyday grind of his working life. McAlpine notes how in many of Clare's Northborough sonnets, "it can seem as though the landscape itself . . . is seeing the scene and taking part in its own description" (90). By these means, Clare intensifies his observation to the extent that he attempts to become "thinged" by the flowers, birds, and grasses surrounding him; in other words, nature is not "at bay" (McAlpine 79) but a source of nourishment, solace, and refuge. To observe so closely that there is almost no distinction between what is seen and who is seeing—a process that can produce microscopically descriptive poetry—reorders "things" in their importance. The "life" of things becomes vital to Clare because he sees himself as part of "things" in a natural world that both makes more sense to him and provides greater "wonder," as McAlpine acknowledges (79). Then again, Clare had been constructively criticized by his friend and fellow poet George Darley, as well as his publisher John Taylor, for making poems that sometimes resembled lists, only interesting to those who wanted the information (Bate 367–369).

However, the intimacy displayed in many of Clare's "nest" poems from the late 1820s reminds us that there is an "animating presence" (Bate 369) in these poems which includes nameable people as well as nonhuman nature. Clare loved to walk by himself but he also enjoyed the company of his local botanizing friends on field walks, such as Joseph Henderson and Edmund Artis, and often included his beloved children. When in good health and good company, Clare would wish to share his world with everyone who came within his poetic or actual territory, to sing with them, talk, idle an hour away, hunt for ferns, flowers, and shells. Yet the nightingale in "The Nightingale's Nest" "raised a plaintive note of danger nigh," in "choaking fear" that her song "might betray her home" (lines 58–61), a subtle allusion to Clare's exposed situation as a labouring-class poet. And in "The Yellowhammer's Nest," just like a "noisome weed that burthens every soil" (line 24), so do snakes enter Eden, "watch such nests & seize the helpless young . . . leaving a housless-home a ruined nest" (lines 26–28). These poems skilfully combine immediacy, drama, and literary knowledge with local detail and bird-lore, yet underlying a descriptive nest poem is another related to Eden's loss and worldly misfortune. Home was always a threatened place for poor labourers; the "housless-home" is terrifying for Clare, and a mark of subsistence living, as Wordsworth noted at Tintern Abbey with its "vagrant dwellers in the houseless woods" (line 21).

Wordsworth's more enigmatic engagement with material nature is determined to find a place within the nonhuman world that he can call "home." His poetry "reflects a struggle to ground . . . imagining in an awareness of the materiality of nature," argues Onno Oerlemans, "the impenetrable reality of surfaces and appearances" (35). In "The Tables Turned," he adds another dimension to "things," urging his readers, to "Come forth into the light" of them (line 15), an immersive approach to matter which may be interpreted in the sense of weightlessness as well as brilliance. In "A slumber did my spirit seal," for example, human is reabsorbed into the landscape, surmounting fear itself yet remaining witness to "earth's diurnal course / With rocks and stones and trees" (lines 7–8). The degree of wildness here, which otherwise might go unnoticed, is in keeping with a project of "contemplative ecopoetics" (*Reclaiming Romanticism* 24), as Kate Rigby terms it, in which the earth's diurnal course is a process of "thinging," in Heidegger's sense of "gathering"—things that are not rocks and stones and trees but spirits (174–182). An attempt to penetrate surfaces and appearances often results in a distinct sense of unease, as Wordsworth's almost obtuse battering creates uncanny manifestations rather than the real thing. In "We are Seven," the poet's insistence on counting children rather than listening to the story disturbs his confidence in facts, while the children are at home—those under the grass by the house, alongside those above the grass where the "little Maid" (line 29) sits, sews, and sings (lines 41–44). We are left with the impression that Wordsworth's acceptance of the child's sums will eventually bring him more comfort than confusion.

David Simpson argues that Wordsworth developed a "poetic persona," the character of "long-distance wanderer across the hills and along the public roads," one of "exemplary loneliness," in keeping with his status as poetic witness to human and nonhuman encounter (54). It is a dangerous, uncomfortable position, in which "[m]otion itself becomes the governing principle, as if the narrator is in a constant state of circulation, a man without home or shelter and worldly destination, a figure of implacable homelessness" (Simpson 56). Yet his vision of home is regularly confirmed by going away—maybe it can only be realized in expanding and contracting cycles of movement in which "spots of time" (*The Prelude* XI.258) offer resting spaces.

Clare's home in the nonhuman world, however, is the only one he knows to be reliable, an earthly blessing that could provide for himself and his family—bean fields, water-meadows, hay-making, sheep tending. All these places and activities were integral to commons and heaths, and were part of a community of nature, human and nonhuman combined. Clare's material involvements are not the "contemplative ecopoetics" (24) that Rigby applies to Wordsworth's encounters with things, whether physical or immaterial, but touchstones of solidity, of stabilizing security in nonhuman everyday reality. For both poets, their attempt to find a refuge in wildness is risky because the vortex of material is not unifying, as Wordsworth would like to persuade Nature to be, but constantly destabilizing. Clare is entangled in matter; his poetry is "forged in relationships with other beings" (Kövesi 100), animal, vegetal, and mineral. Clare so levels his sights to the things around him that he willingly blends his materiality with theirs, "hid in the shadows of the meadow hay" with flies "brustling the rustling rushes" in their delight ("A Flye," lines 9, 6). He develops a poetry that permits him a literary place he aspires to, which also allows poetic freedom to celebrate pragmatic and poetic "wilding." Clare's re-creation of birds, plants, and animals in poetic form rewilds them into a freedom he wished for himself: it is a community of freedoms, combining nonhuman and human knowledges. Rewilding his beloved landscapes allows a self-wilding of John Clare; he becomes "Free as spring clouds and wild as summer flowers" ("The Mores," line 16). Yet Clare's "wild" is also founded on that quality of mental and physical freedom to choose that he needs to negotiate each day in order to find some equanimity with his lack of choice. Wild beings have choices that domestic ones do not; therefore, wildness is something to be cultivated—a textual and material enigma.

In order to create his place in the written world, Clare needed to establish a material world that tallied with the images projected by poets with greater access to choice. His spatial constrictions appear to offer prospects of actual wildness that are unavailable to Wordsworth, Coleridge, and Southey, in spite of their surroundings in Somerset and the Lake District, because of Clare's particular involvement in a domestic as well as wild nonhuman landscape. Wild things include wild acts, especially when lack of freedom creates what might be called "break-out" impulses. When he was finally able to obtain a copy of James Thomson's *The Seasons*, Clare became guilty of appropriating civilized values and could only "go even further down the primrose path, by making a literal, land trespass" over the wall of Burghley Park, home of one of his eventual patrons, the Marquis of Exeter, as John Goodridge and R. K. R. Thornton have extensively discussed (16). They liken this episode to Wordsworth's boat-stealing transgression in childhood, depicted in Book I of *The Prelude*: "It was an act of stealth / And troubled pleasure," yet formative, for "not without the voice / Of mountain-echoes did my Boat move on" (lines 388–390). This was the same voice that Wordsworth heard throughout his youth and growth into a poet with "Both pain and fear, until we recognize / A grandeur in the beatings of the heart" (lines 440–441). The paradox here is that while Wordsworth was confronted by elements that might genuinely be considered part of the "wilderness" of Cumbria, and his act was one of "wildness" in that it was an antisocial act, Clare is trespassing *into* the establishment away from the very wildness he eulogizes. His formative act is, indeed, transgressive, and the amount of manoeuvring necessary to gain a copy of *The Seasons* demonstrates how working people were effectively barred from gaining access to privileged texts. Clare bribed another boy to tend the horses, and spent a working morning reading, in someone else's grounds—a "three-fold trespass on the time, culture and land of his social superiors" (Goodridge and Thornton 13). Bribery, shirking, and trespass: these would be considered rebellious, wild acts, if Clare had been caught.

Not only were there barriers to prevent theft of wild animals owned by those with greater privilege—pheasants, rabbits, or deer—but the very presence of the labouring poor on what looked like wild land became an excuse to harass anyone deemed out of place. Pain and fear were more immediately threatening for Clare: like the nightingale with "choaking fear," he had reason to be on his guard and silent. In Wittering Heath, another part of the Marquis's estate behind another wall, Clare was asleep after one of his rambles, waking to hear the keepers and their dogs. Idleness was suspicious in a young labourer "and I shoud have been taken up as a poacher undoubtedly" if they had chosen to do so, writes Clare (*By Himself* 100). His "night walking" was "associate with the gipseys robbing the woods of the hares and pheasants" because he "was often in their company," as he records in his "Autobiographical Fragments" (78). Clare represents them as sharing the characteristics of the wildlife they depended on: secretive, impulsive scavengers, always ready to move on. Their attraction was that they "know the woods & every foxes den," he writes, "the rabbits know them & are almost tame" ("The gipsies," lines 9, 12). Clare emulated and learned from their fiddle-playing and envied their spirited, peripatetic ways of living which were "tickling temptations to [his] fancy" (86). The potential for licenced vagrancy appealed with that same sense that a child might wish to join the circus. Being "ignorant in the ways of the world," as Clare notes in his "Autobiographical Fragments," gives them a child-like charm; their

potential for "loose . . . morals" offered an enviable wildness (83), yet their freedom was more imagined than actual in a regulated society.

In an early poem, Clare writes from the gipsy perspective: "The Gipseys life is a merry life" in which they "live untythed & free" ("The Gipseys life," lines 1, 4). Later, in the asylum in Epping Forest, Clare writes with a different sense of what it is to be trapped in "The Gipsy Camp," where the man "seeks his squalid camp, half hid in snow, / Beneath the oak, which breaks away the wind" (lines 5–6). Clare is witness to their poor food and meagre lives, yet lets them be the beings they are: "'Tis thus they live—a picture to the place; / A quiet, pilfering, unprotected race" (lines 13–14). He makes ironical reference to their "picturesque" status, knowing that "[t]he picturesque view of the gipsy camp was always from a distance. Clare was the only English writer of the period to be entertained in a gipsy camp, and he gives us insights from close quarters" (Lamont 24). Unlike Wordsworth and William Cowper, who were both uncomfortable about the apparently indolent nature of gipsy life, Clare knew what gipsies did at first hand; they were fortune-tellers and horse-traders, itinerant field workers and chair menders; they made and sold things. Objectified at arm's length, they were viewed as feral, like dogs and cats grown too wild to be admitted to respectable homes, yet simultaneously dependant and suspicious. Sarah Houghton-Walker notes how "unreadable" Cowper found the gipsy camp he depicts in *The Task* (661); similarly, James Garrett comments on how Wordsworth finds his "Gipsies" "uncounted and unaccountable" (619) in his imaginative system (a little like the children in "We are Seven"). They are an "unbroken knot" (line 1), unmoved by the light of things—whether by the sun, "outshining like a visible God" (lines 13–15), or by the moon, who "looks" at them but they "regard not her" (lines 20–21)—themselves lit only by red fire, the "colouring of night" (line 6). That Clare had no difficulty in "reading" gipsies is relevant in a culture that was still largely oral, especially in Clare's youth. His ambivalent tone through successive poems and prose pieces on gipsy life—vacillating from "happy boys" ("The Gipseys life," line 2) to an "unprotected race" ("The Gipsy camp," line 14)—shows how these "natural allies," as Goodridge and Thornton describe them (39), could also be establishment enemies, his establishment being poetry. "Natural allies" represents a good compression of the virtues Clare sought in non-human Otherness: a togetherness that respected natural bounds and that benefited from mutual education in nature's mysteries. Dobbin the horse learned his "tricks" (line 145) of trespass from the gipsies in "Going to the fair," and "felt no shame" (line 152), an example of "gipsies and animals together finding paths to freedom" (Goodridge and Thornton 44). For establishment poets, though, gipsies were "wild things" of an uncanny order. Their wandering habits affect Wordsworth and Clare with different levels of anxiety; Clare could end up in prison for his own wandering, let alone for spending time with other wanderers. Wordsworth fears that his wandering, such as it is, will be taken for idleness, which might account for his persistent

interrogations; they give wandering a purpose, as though he were a roving reporter.

Clare's strange and opportunistic encounters can be as vagrant as any of Wordsworth's wanderers, embedded in material involvement, yet opening out into uncanny quests for wider freedom of understanding. "The Mouse's Nest" is a good example of Clare's struggle to unite exterior explanation with interior encounter at a pragmatic level, becoming an expression of "the light of things." Clare expands closely observed detail into a moment of universal significance, a world in miniature:

I found a ball of grass among the hay
& proged it as I passed & went away
& when I looked I fancied somthing stirred
& turned agen and hoped to catch the bird
When out an old mouse bolted in the wheat
With all her young ones hanging at her teats
She looked so odd & so grotesque to me
I ran & wondered what the thing could be
& pushed the knapweed bunches where I stood
When the mouse hurried from the crawling brood
The young ones squeaked & when I went away
She found her nest again among the hay
The water oer the pebbles scarce could run
& broad old sexpools glittered in the sun. (lines 1–14)

Here, a domestic scene—farmyard, lane, field—is transformed into an exercise in understanding freedom and choice. There is a sense of brutality—Clare was after a bird and he "proged" an unassuming ball of grass. Yet, if we substitute an "old man travelling" for the grass, we are not so far from that disinterested compassion that Wordsworth exhibits at his best. Sarah Houghton notes Clare's inclusivity: he is "only ever one more aspect of the natural world, never superior to it . . . [A]ll features, even when they are understood to be insentient, are parts of his world as important as his fellow men" (787). There is also an implicit awareness that the mouse has not allowed this encounter, and now Clare must acknowledge his responsibility for it. He poetically returns the mouse to her nest, using the same phrase, "among the hay," re-placing himself in juxtaposition to the event, not as part of it.

Clare offers what is close to a simultaneous narrative, "one complex manifold of simultaneous impressions" (Barrell 157) in which the whole is viewed all at once in a series of compartments, like a Renaissance altarpiece. It is not possible to do more than step with him, passing, poking, going away, coming back, running towards, going away, like some curious country dance. Each movement precipitates increased knowledge without any explanation, only observation. And the environment is a picture of "co-dependency"

in which the elements "presuppose what is not of our making," as David Cooper argues for the "epiphanic" value of gardening (146). Clare invites an intimate connection here because he appears to be content to let disparate incidents create their own harmony—or discord. Each element is there "vividly and saliently" in front of you "just as it is," writes Cooper concerning garden phenomena like grass, water, stones (148). "If ever a poet wrote with his eye on the object," comment his editors, "it is Clare in the Northborough poems" (Robinson, Powell and Dawson V: xxiv). Except that objects are "completed," whereas "things" are "gatherings of materials in movement" (Ingold 437). They avoid completion in favour of entanglement. So we have a finishing couplet that apparently bears no relation to the poem as a whole, and yet expands the poem as a whole in a similar manner to the expansion of the turning world in "A slumber did my spirit seal." Here, the world turns on things and how they are lit: "The water oer the pebbles scarce could run / & broad old sexpools glittered in the sun."

Angus Fletcher acknowledges the "sense of the suddenly strange" in the poem's final couplet (60–61). However bright the sun, it glitters, it does not warm; the stream—often the only source of water for human and non-human—has almost vanished. By association, the final couplet echoes rural precariousness, subsistence dwelling for all life. "Responsibility is not a calculation to be performed," observes Barad; "it is a relation always-already integral to the world's ongoing intra-active becoming and not-becoming" ("Quantum Entanglements" 265). On the point of choice, Clare admits a sense of inadequacy, a need to step back, and in doing so he allows the cycle of "becoming and not-becoming" to be maintained. He cannot let go of the image of destruction, though, and is almost absorbed into the sonnet's final couplet in his desire to retreat from interpretation. "The Mouse's Nest" creates a feral scene, not decorative pastoral where "wild" flowers adorn a rustic gate; there is a combined sense of random brutality and confusion, synthesized in the concluding couplet into what Kate Rigby has termed a "poetics of negativity, in which it is not the adequacy of the poetic word but its perpetual falling short that directs us toward an earth and a world beyond the page" (*Topographies* 12). Clare has no words to express the parallel senses of imprisonment and freedom displayed by mouse family and himself as combined victims and catalysts for insight. All are levelled, such that they can no longer "run."

Sarah Houghton considers Wordsworth's more elevated view and prospect, in every sense, as problematic in attempting to create a poetry of community with Others. However much Wordsworth despairs at human ability to create a better world for all, he is still wedded to the hierarchical system that sees humankind at the top.

> Human society remains separate from and dominant within the community of nature, and Wordsworth acknowledges that the 'many into one incorporate' remains a product of the exercise of the poetic

> imagination. Wordsworth thus frames a hierarchy of community, which undoubtedly contains within it the brute multitude, and the protective, enchanting leaves and stars, but in which Man is both the pinnacle and the enabling condition of the structure. (Houghton 785)

By contrast, many of Clare's poems are accounts of reciprocity, human and nonhuman combined, as seen in "The Pasture," from "the swain at his plough," (line 2) to cow-herding, shepherding, "maid[s]" weeding (line 25), and finally "in harvest," when "folks lay / Underneath the green shade of a tree" (lines 33–34). Clare's vision of natural freedom is defined not by mountains and waterfalls but by the interaction of human and nonhuman with the "life of things"—in this case, traced through the perspective of mole hills. Moles were symbols of freedom, resistance fighters, slain by enclosers who "hung the moles for traitors" ("Remembrances" 69). Katey Castellano considers the mole-catcher's role in spreading his "commoning knowledge" (166), yet the knowledge of commons is better understood by the moles themselves, underground and ubiquitous, "creating spaces for the thriving . . . of other species" (173), a communitarian reciprocity.

From the first verse of "The Pasture," in which the scene is set like a De Wint painting, the peewit leads us to the plough, to the cows pressing down the grass, and to the mole who "roots her hillocks anew" (line 5) and thus across the field "from hillock to hill" (line 18). In the foreground, Clare draws us to look at the "cushions" (line 10), resting places for the weary, then allows the mole-hills to open out the prospect: "And see how they checker the plain / And the old hills swell out in the sun" (lines 12–13). The shepherd uses them for pillows, the weeding maid picks and chooses her seat, the haymakers find the hills make perfect picnic tables. This is an investment in wildlife, a cooperative venture whereby moles literally support the wider community of human and nonhuman alike. In particular, the moles' little heaps of earth (which Clare gathered to improve the soil in his garden) provide rest; they allow that space to dream, even for a few moments; they are reasons for stopping; they frame the story. The moles and their subversive workings embody Wordsworth's question to his friend in "Expostulation and Reply":

> 'Think you, 'mid all this mighty sum
> Of things for ever speaking,
> That nothing of itself will come,
> But we must still be seeking?' (lines 25–28)

By these means, Clare puts himself into the combined associations of human and nonhuman experience and knowledge, recognizing that the "wild" knowledge that humans can absorb without "proging" creates both wild and domestic freedom. In this was Clare's freedom: no one could take away this aspect of encounter with the life of things, although it was curtailed by restrictive, often punitive systems. Wordsworth, however disruptive his poetics turn out to be on close reading, aims for a harmony of purpose in the life and light of things. His "Old man travelling" is at one with the "animal tranquillity and decay" of the poem's title, in a blank verse form so pared down it is as though the "little hedge-row birds" (line 1) pecked it out. The title implies that decay is as much a part of animal tranquillity as it is of life, an evenness of temper: "Long patience has such mild composure given, / That patience now doth seem a thing, of which / He hath no need" (lines 10–12). As a "thing," patience takes on tangibility, evident in the combined behaviours of the old man and the little birds: they are "patience," embodied, materialized. Patience as a virtue has lost any meaning, because that implies activity beyond what is needful for the present moment alone. The title also risks aligning human and bird, reducing supremacy of either, but also reducing the status of human to "wild." Thus, Wordsworth as witness cannot resist proging the human; he becomes the interlocutor, wishing to probe where the old man is going and why. The reply is shocking in its contrast: "I am going . . . to take / A last leave of my son, a mariner," who is dying as the result of a sea-battle. Timothy Morton comments on how this interrogation turns a poem of natural reciprocity between beings into a war poem: "something apparently simple, inevitable, and obvious becomes strange, intimate, and painful" (49). There is a new wildness in the disharmony offered here, without any extra commentary, similar to that present in "A slumber did my spirit seal," and also evident in the clash of domestic circumstances that Clare records in "The Mouse's Nest." Encounter becomes vibrant, filled with the life (and death), the light (and darkness) of things.

These encounters are not wild in the traditional sense of that term. There are no monsters, not even in Wordsworth's early terror that the cliff was coming after him for boat-stealing. The monsters are internal, for both poets. Ultimately, Wordsworth retreated from nature's "infinite multitude of indeterminate beings," having failed to purify "things" by reducing their materiality (Barad, "On Touching" 218). As Tim Fulford comments: "To love nature, Wordsworth shows, involves remaking it in our own image," complete with the "traditional hierarchies and inequalities" that were painful for Clare and his community of things (26). Even in his most famous poem, "I Am," Clare is "level with the grass," as Simon Kövesi points out (95). Clare claimed, in another asylum poem, "Sighing for Retirement," that he "found the poems in the fields / And only wrote them down" (lines 15–16), a familiar plea from faux-naïve itinerant ballad-makers. But more intriguing is the line "God bade me make my dwelling there" (11): Clare is not only the messenger of fields but blessed by the fields' Maker to be so. Kövesi considers that Clare is "conventionally" redeemed through his address to God, in this poem and "I Am" (148); however, twenty-first-century constructions of earlier religious adherence can be problematic. It is Clare's intense focus on what "matters" that has a redemptive effect. Here is the spirit, here is the thing, complementary opposites that resolve into truths, water that cannot run, sun that does not warm. In "Sighing for Retirement," Clare finally rests in the knowledge that Wordsworth resisted: there is no hierarchy in nonhuman entanglement; there is only the extended knowledge of continual encounters, revolving in the almost-wild, the uncounted and unaccountable interactions between one thing and another.

Works cited

Ashbery, John. *Other Traditions.* Harvard UP, 2000.

Barad, Karen. "Interview with Karen Barad." *New Materialism: Interviews and Cartographies.* Ed. Rick Dolphin and Iris Van der Tuin. Open Humanities P, 2012, pp. 48–70.

———. "On Touching – The Inhuman that Therefore I Am." *Differences*, vol. 23, no. 3, 2012, pp. 206–223.

———. "Quantum Entanglements and Hauntological Relations of Inheritance: Dis/continuities, SpaceTime Enfoldings, and Justice-to-Come." *Derrida Today*, vol. 3, no. 2, 2010, pp. 240–268.

Barrell, John. *The Idea of Landscape and the Sense of Place 1730–1840.* Cambridge UP, 1972, 2000.

Bate, Jonathan. *John Clare: A Biography.* Picador, 2003.

Bloomfield, Robert. "The Farmer's Boy." *Selected Poems: Robert Bloomfield*, edited by John Goodridge and John Lucas. Trent Editions, 1998, pp. 1–43.

Castellano, Katey. "Moles, Molehills, and Common Right in John Clare's Poetry." *Studies in Romanticism*, vol. 56, no. 2, 2017, pp. 157–176.

Clare, John. "A Flye." *Poems of the Middle Period*, 5 vols, edited by Eric Robinson, David Powell and P. M. S. Dawson. Oxford UP, 1996–2003, V, p. 408.

———. "The Gipseys Life." *Poems of the Middle Period*, 5 vols. Ed. Eric Robinson, David Powell and P. M. S. Dawson. Oxford UP, 1996–2003, IV, p. 52.

———. "The Gipsies." *Poems of the Middle Period*, 5 vols. Ed. Eric Robinson, David Powell and P. M. S. Dawson. Oxford UP, 1996–2003, V, p. 375.

———. "The Gipsy Camp." *The Later Poems of John Clare*, 2 vols. Ed. Eric Robinson and David Powell. Oxford UP, 1984, I, p. 29.

———. "Going to the Fair." *Poems of the Middle Period*, 5 vols, edited by Eric Robinson, David Powell and P. M. S. Dawson. Oxford UP, 1996–2003, III, pp. 91–118.

———. "Hay Making." *Poems of the Middle Period*, 5 vols, edited by Eric Robinson, David Powell and P. M. S. Dawson. Oxford UP, 1996–2003, IV, p. 218.

———. "I Am." *The Later Poems of John Clare*, 2 vols, edited by Eric Robinson and David Powell. Oxford UP, 1984, I, p. 396.

———. "The Mores." *Poems of the Middle Period*, 5 vols. Ed. Eric Robinson, David Powell and P. M. S. Dawson. Oxford UP, 1996–2003, II, pp. 347–349.

———. "The Mouse's Nest." *Poems of the Middle Period*, 5 vols. Ed. Eric Robinson, David Powell and P. M. S. Dawson. Oxford UP, 1996–2003, V, p. 246.

———. "The Nightingale's Nest." *Poems of the Middle Period*, 5 vols. Ed. Eric Robinson, David Powell and P. M. S. Dawson. Oxford UP, 1996–2003, III, pp. 456–461.

———. "The Pasture." *Poems of the Middle Period*, 5 vols. Ed. Eric Robinson, David Powell and P. M. S. Dawson. Oxford UP, 1996–2003, IV, pp. 580–581.

———. "The Primrose." *The Early Poems of John Clare*, 2 vols. Ed. Eric Robinson and David Powell. Oxford UP, 1989, I, p. 182.

———. "Remembrances." *Poems of the Middle Period*, 5 vols, edited by Eric Robinson, David Powell and P. M. S. Dawson. Oxford UP, 1996–2003, IV, pp. 130–133.

———. "Sighing for Retirement." *The Later Poems of John Clare*, 2 vols, edited by Eric Robinson and David Powell. Oxford UP, 1984, I, p. 19.

———. "The Yellowhammer's Nest." *Poems of the Middle Period*, 5 vols. Ed. Eric Robinson, David Powell, and P. M. S. Dawson. Oxford UP, 1996–2003, III, pp. 515–517.

Clare, John. *John Clare By Himself*, edited by Eric Robinson and David Powell. Carcanet, 2002.

Cooper, David E. *A Philosophy of Gardens*. Clarendon P, 2006.

Fletcher, Angus. *A New Theory for American Poetry: Democracy, the Environment, and the Future of the Imagination*. Harvard UP, 2004.

Fulford, Tim. "Wordsworth's 'The Haunted Tree' and the Sexual Politics of Landscape." *Romantic Circles: Romanticism and Ecology*, 2001, pp. 1–26, www.romantic-circles.org/praxis/ecology/fulford/fulford.html.

Garrett, James M. "The Unaccountable 'Knot' of Wordsworth's 'Gipsies'." *Studies in English Literature 1500–1900*, vol. 40, no. 4, 2000, pp. 603–620.

Gilchrist, Octavius. "Some Account of John Clare." *The London Magazine*, 1820, pp. 7–11.

Goodridge, John. "John Clare and Eighteenth-century Poetry: Pomfret, Cunningham, Bloomfield." *The Eighteenth Century*, vol. 42, no. 3, 2001, pp. 264–278.

Goodridge, John and R. K. R. Thornton. *John Clare, The Trespasser*. Five Leaves Publications, 2016.

Heidegger, Martin. "The Thing." *Poetry, Language, Thought*, Translated by Albert Hofstadter. Harper Collins, 1971, pp. 174–182.

Houghton, Sarah. "The 'Community' of John Clare's Helpston." *Studies in English Literature, 1500–1900*, vol. 46, no. 4, 2006, pp. 781–802.

Houghton-Walker, Sarah. "William Cowper's Gypsies." *Studies in English Literature, 1500–1900*, vol. 48, no. 3, 2008, pp. 653–676.

Ingold, Tim. 'Towards an Ecology of Materials'. *Annual Review of Anthropology*, vol. 41, 2012, pp. 427–442.

Kövesi, Simon. *John Clare: Nature, Criticism and History*. Palgrave Macmillan, 2017.

Lamont, Claire. "John Clare and the Gipsies." *John Clare Society Journal*, vol. 13, 1994, pp. 19–32.

Mahood, Molly. *The Poet as Botanist*. Cambridge UP, 2008.

Marshall, William Humphrey. *Planting and Ornamental Gardening: A Practical Treatise*. J. Dodsley, 1785.

McAlpine, Erica. "Keeping Nature at Bay: John Clare's Poetry of Wonder." *Studies in Romanticism*, vol. 50, no. 1, 2011, pp. 79–104.

McEathron, Scott. "*Lyrical Ballads*, and the Problem of Peasant Poetry." *Nineteenth-Century Literature*, vol. 54, no. 1, 1999, pp. 1–26.

Morton, Timothy. *The Ecological Thought*. Harvard UP, 2010.

Oerlemans, Onno. *Romanticism and the Materiality of Nature*. U of Toronto P, 2002.

Oppermann, Serpil. "From Ecological Postmodernism to Material Ecocriticism: Creative Materiality and Narrative Agency." *Material Ecocriticism*, edited by Serenella Iovino and Serpil Oppermann. Indiana UP, 2014, pp. 21–36.

Potkay, Adam. "Wordsworth and the Ethics of Things." *PMLA*, vol. 123, no. 2, 2008, pp. 390–404.

Rigby, Kate. *Reclaiming Romanticism: Towards an Ecopoetics of Decolonization.* Bloomsbury, 2020.

———. *Topographies of the Sacred: The Poetics of Place in European Romanticism.* U of Virginia P, 2004.

Simpson, David. *Wordsworth's Historical Imagination: The Poetry of Displacement.* Methuen, 1987.

White, Simon J. *Romanticism and the Rural Community.* Palgrave Macmillan, 2013.

Wordsworth, William. "Expostulation and Reply." *William Wordsworth: Major Works*, edited by Stephen Gill. Oxford UP, 1984; revised, 2000, pp. 129–130.

———. "Gipsies." *William Wordsworth: Major Works.* Ed. Stephen Gill. Oxford UP, 1984; revised, 2000, p. 332.

———. *Home at Grasmere. William Wordsworth: Major Works*, edited by Stephen Gill. Oxford UP, 1984; revised, 2000, pp. 174–199.

———. "Lines written a few miles above Tintern Abbey." *William Wordsworth: Major Works*, edited by Stephen Gill. Oxford UP, 1984; revised, 2000, pp. 131–135.

———. "Old Man Travelling." *William Wordsworth: Major Works*, edited by Stephen Gill. Oxford UP, 1984; revised, 2000, p. 29.

———. *The Prelude. William Wordsworth: Major Works*, edited by Stephen Gill. Oxford UP, 1984; revised, 2000, pp. 375–590.

———. "A Slumber Did My Spirit Seal." *William Wordsworth: Major Works*, edited by Stephen Gill. Oxford UP, 1984; revised, 2000, p. 147.

———. "The Tables Turned." pp. 130–131.

———. "We Are Seven." *William Wordsworth: Major Works*, edited by Stephen Gill. Oxford UP, 1984; revised, 2000, pp. 83–85.

3 Plumbing the depths of wildness

From the picturesque to John Clare

Markus Poetzsch

As a socially disinterested theory of art and visual pleasure that inspired a touristic approach to the natural world, the picturesque has long been a target of environmentalists and ecocritics who associate its commodification of nature with anthropocentrism. While its originators, as Kevin Hutchings argues, were "frequently motivated by a love of nature" (179) and their theories and guidebooks proffered a broad invitation to readers to find meaning in the environment as well as in products of material culture, the legacy of the picturesque is complicated, on one hand, by its elision of land and landscape, and by its separation, on the other, of subject and object in human-natural relations. Yet even as scholars debate the shifting valences of these organizing binaries of environmental thought, the picturesque continues to exert considerable influence on our relationship to nature—in part, as Allen Carlson (137) observes, because aesthetic appreciation is fundamental to how we regard and engage the world. We protect and defend not only what we love, to borrow Wendell Berry's formulation (86), but also what is aesthetically pleasing, what stimulates our curiosity and satisfies our appetite for novelty. This is perhaps especially true of environments that have escaped the incursions of human activity and thus confront us with notions of what is primeval, untamed, other, wild.

Indeed, as Isis Brook has suggested, it is the picturesque vogue for wildness that informs our "current interest in . . . protecting pristine wilderness or traditional rural landscapes" (105). One of a growing number of critical reassessments of the picturesque that includes recent publications by Vittoria Di Palma, Roger Paden, and Essaka Joshua, Brook's essay builds upon the work of Eugene Hargrove who advanced provocative links between Romantic-era aesthetics and contemporary programmes of wildlife preservation. Focussing specifically on garden design, Brook characterizes the picturesque as a "misunderstood" aesthetic model and proceeds to isolate six of its governing features or "themes" (variety, intricacy, engagement, time, chance, and transition), the sum of which, she contends, may be conducive to fostering "a love of real wild nature" (106). Of especial interest in this formulation (aside from the contentious conjunction of modifiers hinting at a transcendent signified) is the alignment of wildness with gardening: what some readers might regard as paradoxical, given the implied element of human superintendence but what also reflects the original theoretical impulses of picturesque theory as articulated by William Gilpin and Uvedale Price. Their work, as I will suggest, inaugurates a particular understanding of wildness as an environment of limited scope and engaging visual forms that is rendered manageable by discrete acts of human mediation. Or as Brook phrases it, "At the heart of the picturesque is a love of wild nature in a small compass" (112).

This idea of contained or framed wildness—always small enough to be manageable—betrays considerable anxiety about what lies beyond the precincts of civilized life and whether that something, that otherness, is at all compatible with human interests or, more fundamentally, human being. The 1790s certainly stoked fears about what "wildness" in the political realm might engender but works such as C.F. Volney's *Ruins* (1796), George Cuvier's lectures on paleontology (1796), and Thomas Malthus's *An Essay on the Principle of Population* (1798) also contributed to a growing apprehension about the stability, one might even say the inevitability, of human civilization. For Romantic-era audiences, as Hutchings points out, the implications of these ground-breaking scientific insights included "the notion that *Homo sapiens* was itself subject to ecological limits, that humans—despite their perceived status as privileged lords of earthly creation—were not immune to the possibility of future extinction" (176). Whether extinction in this context was understood by eighteenth- and nineteenth-century readers exclusively as species eradication or whether it implied what Jacques Khalip qualifies as "a revision of 'life' under other terms" (629) is not entirely clear, but the effects of such existential provocations even on disciplines beyond the scientific cannot be underestimated. Indeed, for connoisseurs of the picturesque the wild became a multivalenced aesthetic construct through which one could conjure not only the soothing nostalgia of "an old order of rural paternalism" (70), as Ann Bermingham argues, but also unnerving glimpses of a dystopian future in which human life is afforded neither environmental priority nor security.

While this anxious, bifurcated response to wildness and a consequent reliance on technologies of mediation to manage such environments may constitute the defining aesthetic attitude of many Romantic writers and artists, the period also witnessed the emergence of more ecologically sensitive approaches to natural unruliness and disorder. Of particular relevance is the work of John Clare—himself "a great poet of mess" (1), as Mina Gorji asserts—who established in his writings a telling counterpoint to the narrative of human dominion that underlies, as well as the interventionist protocols that overarch, the picturesque. Clare's understanding of wildness is arguably the most modern of Romantic perspectives because it accounts not only for the reality of diminishing wilderness spaces but, equally, for the limitations of an anthropocentric perspective in mediating the world of nature. Notably, what Clare knew of the wild as physical space was informed not by travel or tourism, the mode of aesthetic enrichment recommended by picturesque theorists, but instead by the relatively narrow precincts of home—the topography of Helpston before and after enclosure. And even this knowledge, as Simon Kövesi argues, was further subject to the inexorable pressures of class, that "social sediment on top of him" (40) by which Clare's leisure time and aesthetic choices were to some extent constrained. The notion of wildness that thus emerges in Clare's work, what he renders so provocatively as his "suiting scene" (line 71) in the poem "Cowper Green," is articulated vertically or deep down into the earth rather than horizontally as a prospect view of bounded dimensions. Deep wildness, as I will suggest, is as much a temporal as a physical designation; it expresses itself in layers of organic growth and decline, a non-hierarchical commingling of species, propagating and passing together, untended, as seeds and as bones.

Picturesque wilderness and the human problem

An immersion in the deep wild depends, however, on a clear articulation of picturesque wilderness, what it comprises as well as circumscribes and what, despite those efforts, eludes its fretful protocols of curation. Clare himself, as I will argue, developed his aesthetic theories in response to the picturesque and thus his own engagement with the wild calls attention to the work of Gilpin and Price, in particular their focus on natural unruliness as a strictly formal quality, a two-dimensional flatness or shallowness manageable by the eye alone, and on the place of humans within such landscapes. Notwithstanding the anxiety it provoked, the wild was arguably the central preoccupation of picturesque theorists and artists alike. More than a generalized context of aesthetic mediation or a pleasant backdrop upon which to evoke keener visual pleasures, wild nature, as Larry Kutchen reminds us, was "intrinsic" to the picturesque imaginary (1). Indeed, the defining qualities of a natural prospect that render it "capable of being *illustrated by painting*" (Gilpin, *Three Essays* 3) include elements of the ungovernable, the unpredictable, and the disordered. Gilpin isolates "roughness" and "ruggedness" as features essential to picturesque landscapes and he counsels prospective travellers in search of such scenes to prioritize—or simply imagine—destinations that have been "unexplored" (47). Price, in like manner, designates "roughness," "sudden variation," and "irregularity" as "the most efficient causes of the picturesque" (37), arguing that scenes "occasioned by accident and neglect" (27–28) are most likely to foreground and intermingle these causes in arresting combinations. The shift here from the "unexplored" (*terra incognita*) to the "neglected" (a broad category of terrain that is either completely uncultivated, left as fallow for a season, or already despoiled as wasteland) is suggestive of the breadth of signification to which wildness as a physical environment was subject. Yet if the topography of wildness (then as now) resists definitive expression and categorization, its aesthetic qualities, including its function in a landscape and its effects upon the viewer, were more uniformly articulated by picturesque theorists. In advising travellers to seek "unexplored" habitats, Gilpin emphasizes both the element of surprise that attends such pursuit—a pleasure all its own deriving from "agreeable suspence [sic]"—and the likelihood that wild landscapes, given their varieties of form and intricacies of design, will potentiate that "employment of the mind" which he regards as central to all aesthetic experiences (48). The picturesque mindset has at its core an active or dynamic principle, a keen sensitivity to, and sophisticated protraction of, pleasure. As Gilpin explains, in wild landscapes:

> we are . . . commonly employed in analyzing the *parts of scenes* . . . We examine what would amend the composition; how little is wanting to reduce it to the rules of art; how trifling a circumstance sometimes forms the limit between beauty, and deformity. Or we compare the objects before us with other objects of the same kind:—or perhaps we compare them with the imitations of art. From all these operations of the mind results great amusement. (*Three Essays* 49)

Wildness thus conceived is a bounded relational field of particularized but disordered objects, one as it were encroaching upon another, and the viewer's role, according to Gilpin, is to edit the scene through a process of "minute examination" from which is derived "a compleat idea of an object" (50). By such analytical mediation—even before the scene is transferred onto canvas—we become, as he suggests, "more learned in nature" (50).

Price, while less inclined to muse on the educative effects of an immersion in the wild, corroborates Gilpin's claim that the mind derives considerable stimulation from scenes shaped by "accident and neglect." Unlike the work of landscape improvers who devote "exclusive attention to high polish and flowing lines" (21), neglected nature, operating seemingly at its own discretion, yields both variety and intricacy of form and it is upon these qualities that Price fixes the human emotion of "curiosity," a state of arousal conducive to the unique pleasures of picturesque travel. Whereas the aesthetic experiences of beauty and sublimity culminate, respectively, in conditions of static delight and astonishment, with the subject "unwilling [or unable] to move, almost to think," the picturesque has a precipitating or potentiating influence:

> Those who have the felt the excitement produced by the intricacies of wild romantic mountainous scenes can tell how curiosity, while it prompts us to scale every rocky promontory, to explore every new recess, by its active agency keeps the fibres to their full tone; and thus picturesqueness when mixed with either of the other characters, corrects the languor of beauty, or the tension of sublimity . . . It is the coquetry of nature . . . [B]y its variety, its intricacy, its partial concealments, it excites that active curiosity which gives play to the mind. (88–89)

Like Gilpin before him, Price charts in the progress of the picturesque encounter a complex interchange of energy that appears to originate in the landscape itself—in the intricacies of its component parts which, depending on perspective, alternately reveal and conceal themselves and thus continually break the lines that would guide the eye to a determination of complete forms or silhouettes—while also kindling in the viewer a perception of the mind's "play," its active processing of the incomplete visual stimuli that simultaneously promise and forestall resolution.

Price, notably, encodes this energy in the language of sexual arousal: from "promontory" to "recess," the picturesque traveller luxuriates in the varieties of visual stimulation on offer from the "coquetry of nature." As a metaphor for the power of the wild, Price's sensual landscape registers the desire for human (specifically, male) control but also the fear of proving unable to exert it. Indeed, the qualities by which nature as a visual field leads the traveller on, so to speak, are framed as strategies of allurement that promise not gratification but unabated longing. Wildness always conceals something in its copious display of variety and thereby maintains an aura of impenetrability. Thus, even as the viewer's curiosity is aroused, it can only be maintained or kept "alive," as Price concedes, by being "unsatisfied" (123). The pleasure of the picturesque encounter by which the eye presides and ranges freely over landscapes and gives the impression of human control is, in other words, predicated on the un-pleasure of quenchless desire, a search that never finds. Read in this light, Price's claim that "a certain irritation or stimulus is necessary to the picturesque" (125) foregrounds an unmistakable element of irresolution. Is a wild landscape ultimately irritating or stimulating for the viewer? Or to ask a more pointed question, of what value is the "active agency" (88) of sensual arousal if its culminating gesture is to the limitations of human agency in finding fulfilment in the wild?

While Price's text never addresses these questions directly, the contours of its analysis reveal unsettling glimpses of the precariousness of human life in the landscapes of the wild. In an extended commentary, for example, on Shakespeare's description in *The Merchant of Venice* of the "moonlight sleep[ing] upon yon bank" (5.1.52), Price offers the following corrective:

> Nothing in that line gives any indication what sort of bank that was, but if you fancy it broken and abrupt, the moon might indeed shine, but it could no longer sleep upon it. The same kind of sympathy that takes place in smaller objects, in broken ground, roots, stones thorns, or briars, where a certain degree of difficulty and irritation is common and familiar, seems to continue whatever be the scale. A fall from a great height, as from the side of a precipice, is equally destructive whether the surface upon which you fall be rugged, or plain: yet the imagination would be differently affected by looking down upon an even surface, or on sharp pointed rocks; and some feeling of that kind I believe is always connected, though we may not at all times be conscious of it, with broken and pointed forms. (121)

The image that Shakespeare paints of the natural world reconciled to itself and to the human observer (in this case, Lorenzo) is repurposed by Price and shifted from an aesthetic register of beauty culminating in "stillness and repose" (121) to a nightmare vision of the picturesque that literally shatters the human-natural bond on "sharp pointed rocks." Price's reimagining or "wilding" of the bank, on which the aesthetic transposition depends, is presumably intended to heighten and generalize the scene's appeal. Yet what the viewer gains in terms of physical or cognitive stimulation from a thanatic contemplation of the rough ground that lies at the foot of a great precipice is almost immediately lost in the brevity of the experience, as Price considers the leap that ends in destruction. Here at last the relentless progress of curiosity is halted, yet the deeper questions raised by this passage concern not only the appetitive trajectory of picturesque exploration but also the place of humans within a landscape of "broken ground." Indeed, what Price foregrounds, in what is ostensibly an elaboration of the artful qualities of nature that unconsciously shape our reactions to it, is the more fundamental problem—the tenuousness, indeterminacy, uncertainty—of human being in the wild.

Gilpin's engagement with this issue—what I will call "the human problem"—manifests itself most directly in his advice to would-be painters about how best to compose a scene. Writing at a time when the growing popularity of landscape art had already begun to subordinate human figures to an ornamental role on the canvas—they were, as Malcolm Andrews explains, subsumed under the broad heading of "staffage" and served predominantly to animate the world of nature (25)—Gilpin encourages even stricter protocols of control:

> In *adorning your sketch*, a figure or two may be introduced with propriety. By figures I mean moving objects, as waggons, and boats, as well as cattle, and men. But they should be introduced sparingly. In profusion they are affected. Their chief use is, to mark a road—to break a piece of foreground—to point out the horizon in a sea-view—or to carry off the distance of retiring water by the contrast of a dark sail, not quite so distant, placed before it. But in figures thus designed for the ornament of a sketch, a few slight touches are sufficient. Attempts at finishing offend. (*Three Essays* 77–78)

The position of "men" in the sequence of figures is indicative of Gilpin's general indifference to the aesthetic interest of his own species. Human figures function essentially like any other vertical marks on the canvas—trees, fences, sails, ruins—in that they break the flow of horizontal lines and thus facilitate that continual revision or recalibration of the visual field which is experienced as the mind's "play." The more "finished" or resolved a human figure, the greater the likelihood that it becomes a focal point on the canvas, thus impeding the untrammelled movement of the eye across the landscape.

Gilpin extends his strictures against human ornamentation even to the appearance of common agricultural tools which, as he notes, "too often introduce preciseness, and formality" (iii). His objection to these hints of realism appears to centre not only on their infringement upon artistic licence but also on the incongruence between wild landscapes and the implements of human domestication. Indeed, the declamatory poem on landscape painting that he appends to the *Essays* establishes the need for thematic compatibility between natural topography and human ornament:

> If rocky, wild, and awful be thy views,
> Low arts of husbandry exclude: The spade,
> The plough, the patient angler with his rod,
> Be banished thence; far other guests invite,
> Wild as those scenes themselves, banditti fierce,
> And gypsey-tribes, not merely to adorn,
> But to impress that sentiment more strong,
> Awakened already by the savage-scene. (lines 578–585)

But even here the risk of distracting the viewer with caricatures or otherwise discrepant staffage inclines Gilpin to more severe measures. Calling to mind the example of Claude Lorrain, an artist who regularly, in Gilpin's view, "crowded scenes . . . With forms ill-drawn, ill-chosen, and ill-arranged" (lines 594–595), he finally enjoins the picturesque connoisseur to "sweep [such ornaments] far from our disgusted sight!" (line 598). It were better for the landscape, he insists, to "leave it desert" (line 593).

While suggestive of a fundamental incompatibility between humans and the wild, at least on canvas, Gilpin's enthusiasm for "desert" landscapes proves rather more difficult to sustain when he is confronted by actual semblances of the wild in his travels of Wales and Scotland. Undertaken ostensibly for personal amusement, aesthetic enrichment, and public edification—as examples of how "To criticize the face of a country correctly" (*Observations* vii)—Gilpin's tours reveal telling "disjunctures between Picturesque theory and the practices that are justified under its name" (Copley and Garside 1). His expectation, for example, that wild environments, the unexplored as well as the neglected, offer a dependable variety of picturesque features and adornments is as often disappointed as exceeded by the topography of Wales, particularly in the watershed around Plynlimon. Though "vast, wild and unfurnished"—in other words, ideally suited to the mind's play—the country is "neither adorned with accompaniments, to be a scene of beauty: nor should I suppose from the accounts I have received, that it's dreary wastes could afford such materials, as could form a scene of grandeur [sic]" (*Observations* 76). Given the failure of the aesthetic frameworks of beauty and sublimity, one would expect Gilpin to assert in their place the mediating influence of the picturesque, yet what he discovers instead is a form of wildness unanticipated by theory: the dreary waste, "monstrous without proportions" (76). So unruly and uninspiring is the environment, so aesthetically and humanly unmanageable, that Gilpin turns the erstwhile wish for "desert" landscapes into a plea for further desertion: "As there is nothing therefore in these inhospitable regions to detain you long; and no refreshment to be had, except a draft of pure water from the fountains of the Wye, you will soon be inclined to return to Rhaader" (76). While Gilpin the aesthetician may yearn for unpeopled wilderness to fire his imagination, as a traveller through the unadorned Welsh uplands his longings are rather more terrene and convivial, centring on liquid refreshment and oases of human hospitality like the market town of Rhayader.

His travels through the Scottish Highlands in the summer of 1776 reinforce the tension between theorized or wished-for wildness (and the vexed role of humans therein) and the state of nature as environmental reality. This phrase "the state of nature" is one that Gilpin regularly deploys in the narrative of his journey in order to convey a sense of primal wildness, of nature "totally untouched by art" (*Highlands* 2: 111). What Gilpin means by "art" in this instance does not appear to include the sketches of the aesthetic tourist, though these too may change a landscape over time by drawing more visitors to it; instead, he implicates the "tools" of agricultural development by which humans have "miserably scratched, and injured the face of the globe" (2: 112). The English countryside, he observes, is replete with these blemishes, most starkly visible in the property markers that "break the great flowing lines of nature" and thus "injure the idea of wildness" (2: 111). In Scotland, by contrast, Gilpin is delighted to find that "we rarely met with any of these intersections. All is unbounded" (2: 112). Scotland, in other words, is a blank canvas of wildness upon which the artist may impose his own compositional order. Yet as Gilpin discovers, unbounded prospects coupled with human absence are not in fact reliably conducive to the evocation of pleasure. On the way to Drumlanrig, for example, he finds himself crossing a "wild country" in the waning hours of day—ideal conditions, as he notes, for the pursuit of "picturesque ideas"—yet the scene proves acutely disappointing:

> Often, when mountains, forests, and other grand objects, float before the eye, their sweeping forms, clad in the shade of evening, have a wonderful effect upon the imagination. But here the objects were neither grand, nor amusing. All was one general blot. (2: 77)

Like the "dreary wastes" that he encounters in the shadow of Plynlimon, Scottish wildness takes on inscrutable forms. The "blot," notably, is a mark of disfigurement or deformity as well as an erasure or concealment; it is not so much the antithesis of the picturesque as an intensification of its desired disorder. Thus here, once again, Gilpin is forced to confront the unmanageable extremes of wildness that appear to expunge not only human presence but even the means—in this case, artistic—by which human beings can establish a relation to the wild.

Gilpin's response to unmanageable wildness, indicative of the picturesque movement as a whole, is to contract its scope and emphasize its most accessible features, thereby reasserting, albeit with diminishing effect, the regulating powers of the human eye and ego. Nowhere is this process more clearly delineated than in his response to the country that lies west of Killin towards Tyndrum, along the river Dochart. He designates it inauspiciously as "a wild country, which nature had barely produced; but had done little to adorn" (*Highlands* 1: 171). The implication here is of a wildness that falls almost outside the bounds of nature's genial fecundity—a too-wild topography that is not simply indifferent but demonstrably hostile to human dwelling. As Gilpin laments, "Even the cottage smoking among a few trees, which almost every heath presents, was not here to be found. All was wide, waste and rude" (1: 171). Confronted with a prospect so exceedingly unpropitious, Gilpin resorts to an extreme process of focalization:

> Here and there indeed a mountain-scene fell within the rules of composition. But in general, we had few forms of picturesque beauty, at least in the larger parts. In the smaller, we often found them; in the winding of rivulets, in their rocky beds, and in their little bustling cascades, of which we had a great variety. (1: 172)

For the reader expecting either the languid pleasure of pursuit or a frisson of physical stimulation so typical of picturesque description, this passage proves disillusioning, an echo surely of Gilpin's own reaction. Despite being surrounded by undulating terrain that heaves "like the ocean into ample swells, and subsid[es] into vallies equally magnificent" (1: 172), he is unable to conjure the kind of prospect view that might divide the landscape—order it, in other words—into the typical tripartite structure of foreground, middle distance, and background. His perspective is not merely narrow but chastened, giving the impression of one fixated upon pebbles while mountains are rolling round about him. Indeed, the only redeeming features that he is able to enumerate (rivulets, rocky beds, cascades) are decontextualized and isolated from the landscape as a whole—in much the same way, one might add, as the viewer himself.

Ultimately then, in its grappling with the wild, the picturesque reveals itself to be a framework for human disenchantment, isolation, and inefficacy, even as it purports to control and regulate by a process of "optical hegemony" (Broglio 19). Malcolm Andrews foregrounds this desire for control when he claims that "the [picturesque] tourist wants to discover Nature untouched by man; and yet, when he finds it, he cannot resist the impulse, if only in imagination, to 'improve' it" (3). The work of Gilpin and Price, however, points to a different outcome, not the assertion but rather the illusion of control; not improvement or a solacing of the ego through creative renegotiations of space on canvas, but rather an apperception of incommensurability or lack of relation between the landscape and the self—a gap unbridgeable even for art. The picturesque response to its own failure to manage the wild is demonstrated by Gilpin in the breaking up of unruly prospects into a series of isolated and decontextualized features, not even scenes, that are presented through a radical narrowing of perspective and a superficial investment in distinguishing details. Bermingham characterizes the product of such aesthetic compromises as a "vignette . . . a centralized composition in a shallow space" (85)—a fitting summation of both picturesque text and image, neither of which can adequately account for the reality of wilderness across space and time, its layers and depths, its variable and changing constitution, above all its ecological and relational implications for human beings in the world.

John Clare and deep wildness

As a counterpoint to the Gilpinian rendering of wilderness as shallow space—that is to say, a space curated for the containment of vastness and the management of its inscrutable disorder by segmentation—I would like to consider the work of John Clare, a writer whose engagement with the picturesque approach to landscape aesthetics is rather more complicated than the scholarly consensus suggests. Citing the commentaries of John Lucas and Scott Hess, Adam White argues that "Most critical work has tended to understand the relationship of Clare's poetry to the picturesque as antagonistic" (38) when in fact the breadth of his poetic output across four decades militates against such conclusiveness and reflects, at the very least, a nuanced and considered approach to what was the dominant artistic paradigm of his day. Indeed, as John Barrell reveals in his foundational study of Clare's environmental aesthetics, the poet's technique, tastes, and ideas about art underwent considerable modification and recalibration between the publication of his *Poems Descriptive of Rural Life and Scenery* in 1820 and the miscellaneous asylum poems written between the 1840s and 1860s. In tracing the circuitous development of what he eventually designates an "aesthetic of disorder" (160), Barrell describes a youthful poet whose early work was "thoroughly dependent" (112) on the eighteenth-century tradition of the poetry of melancholy, on the familiar tropes of loco-description inaugurated by James Thomson and, most importantly, on the picturesque model of composing a textual landscape (148). Not surprisingly, the word "picturesque" is deployed in several of Clare's better known poems, such as "Shadows of Taste," "Pleasant Places," and "Pilsgate Hill," and it appears in these texts without any trace of irony or derision. The latter sonnet, for example, finds the speaker stationed on the titular hilltop (about 4 km west of Helpston, above the river Welland) in command of an extensive prospect view:

> On this hill top Ill linger for awhile
> & count the trifles leisure loves to see

Tis always sweet some minutes to beguile
To see what pictures in the scene may be
That will delight . . . (lines 1–5)

Clare's transformation of a familiar environment into a "scene" and his further compression and distillation of that scene into "pictures" speak to the direct influence of the picturesque. The poet goes on to characterize these optical compositions as being "picturesque to me" (line 5), which may suggest an appeal to subjective authority rather than the strict formulas of pleasure bequeathed by Gilpin and Price, yet in the lines that follow a very conventional catalogue of landscape ornaments emerges, including "yonder sweeping bridge at distance seen / & yon rude brook that leads tot crooking free / At natures easy will thro meadows green" (lines 6–8). What structures these descriptive sketches, notably, is the roving, detached eye of the picturesque connoisseur, not the actual movement of the poet through a physical environment, not the intimate and particular—or "micromorphological" (Irvine and Gorji 119)—engagement with the world that readers of Clare have come to expect. No less surprising is his reference to "natures easy will," a phrase which would appear to sanction the sorts of appropriations, both visual and physical, for which picturesque aestheticians and landscape gardeners were known and often maligned. Clare himself, it should be added, was not above such indiscretions, as his lifelong habits of specimen collection, his tendency to transplant local flora, and his invasive engagements with birds' nests and eggs reveal.

My point in tracing Clare's aesthetic development to this controversial juncture is merely to emphasize that he, not unlike Wordsworth who famously likened the picturesque to an "infection" (*Prelude* 11.156), had a complex and ongoing relationship with the dominant artistic modes of his day, so much so that some of his early work hints conclusively at the influence exerted by the picturesque even as other works from the same period signal variances so profound that no degree of scholarly sleight of hand can reconcile them with theories of Price or the touristic strictures of Gilpin. For Barrell, interestingly, Clare's early entrancement with the picturesque does not substantially compromise the poet's "aesthetic of disorder" which, in foregrounding natural particularities at the expense of compositional order, appears to defy the rules of picturesque perspective. Yet Barrell's argumentative line, subsequently retraced by Timothy Brownlow who also emphasizes Clare's "creative disorder" (132), depends on the same rhetoric of irregularity central to picturesque descriptions of the wild. What Barrell and Brownlow value in Clare—his aesthetic or artistic "wildness"—is precisely what Gilpin and Price appreciate in landscapes governed by roughness and unruliness, the difference being that Clare's readers see no reason to edit the poet's choices, while advocates of the picturesque endeavour, through the artifice of composition and perspective, to adjudicate the disorder of the natural world. In both instances, however, the conceptual affiliation of wildness with disorder filters this construct through the lens of human visuospatial perception. In describing Clare's "aesthetic of disorder," for example, Barrell uses the analogy of "a 'continuum' of images [that] will not coalesce into a composition" (156). The emphasis on "images"—what is apprehended by the eye in the here and now—inaugurates merely another version of "shallow space," with Clare's poems, like the sketches or canvases of the picturesque artist, serving as momentary glimpses of wild nature. The natural elements or features described in the poems may not be ordered or composed with the same degree of control as a painted landscape in the picturesque style—indeed, that is what distinguishes their effect, according to Barrell—but they transmit "images" all the same and thus regulate wildness by the protocols of what is visually distinctive or noteworthy. The wild, as such, is a designation based entirely on form, not substance; it is distilled through the immediacy of visual cues—which, depending on one's preference, may be ordered according to rules of picturesque composition and perspective, or experienced as an unregulated, free-form succession of images, such as Barrell describes.

Neither view, as I would like to suggest, represents an adequate characterization of Clare's representation of wildness as deep space and the text that informs my analysis—arguably his most extensive reflection on how wildness integrates but also supersedes the human—is "Cowper Green," a poem published in 1821 in volume I of *The Village Minstrel* and centred ostensibly on a small field southwest of Helpston village, within a stone's throw of Royce Wood to the east and the Roman quarry, Swordy Well, to the south. The name of this tract of land has been variously spelled Cauper Green, Cooper Green, and Cowper Green, the latter eventually chosen at the behest of Clare's publisher, John Taylor, who apparently deemed "Cauper" far too ugly a word for poetry (Bate 54). That Clare would yield to Taylor on this point, given his commitment to preserving local place names and dialect, is perhaps surprising but the episode already signals a tacit understanding on the poet's part that wildness is not to be subjected to or constrained by the authority of human designations; the labels fall off, as it were, as soon as they are affixed and make no meaningful contact with the essence of a place. Then again, essence is the wrong word in this context, because Clare in his descriptions of Cowper Green never arrogates to himself an essentializing perspective by which wildness can be distilled or rendered as a single comprehensible idea. He, like Gilpin, may seize upon small or seemingly manageable environments, yet his investment in such habitats is not a reaction to the perceived disappointments of more sweeping views. On the contrary, Clare is prompted by a deep fascination with the particularities of bounded local space which to other passersby, especially those of his own class, are often overlooked. Indeed, in the midst of his leisurely approach to Cowper Green, at the end of a long workday when "All is silent, free from care" (line 7), he is impelled to circumvent a "selfish clown" (line 10) whose "low [and] vain" (line 13) conversation threatens for a moment to disenchant the prospect of aesthetic immersion. Yet what reorients the poet's focus, almost as quickly as it was diverted, are the first hints of wild nature resolving into distinct shapes, "the insect and the weed [that] / Court [his] eye" (lines 20–21).

In proceeding to describe Cowper Green, to render accurately the native wildness that he familiarizes as "my suiting scene," Clare embarks on a meticulous enumeration of features, textures, impressions, and experiences that establish a direct counterpoint to the cultivated landscapes of the picturesque. As he notes,

> Some may praise the grass-plat whims,
> Which the gard'ner weekly trims;
> And cut-hedge and lawn adore,
> Which his shears have smoothen'd o'er:
> But give me to ponder still
> Nature, when she blooms at will,
> In her kindred taste and joy,
> Wildness and variety;
> Where the furze has leave to wreathe
> Its dark prickles o'er the heath;
> Where the grey-grown hawthorns spread
> Foliag'd houses o'er one's head;
> By the spoiling ax untouch'd
> Where the oak tree, gnarl'd and notch'd,
> Lifts its deep-moss'd furrow'd side,
> In nature's grandeur, nature's pride. (lines 39–54)

This passage signals the peculiarity of the poet's aesthetic preferences in an age ruled, on the one hand, by the shaven lawns and trimmed hedges of the landscape improver, and, on the other hand, by the manicured wildness of the picturesque artist intent on bending the "will" of nature to his own designs. Clare's focus, instead, is devoted entirely to the agency of nature—its (or rather, *her*) free expression—as embodied in "the furze [that] has leave" to grow and spread where it will and in the oak tree, "untouch'd" by the instruments of human spoliation, that supports in its upward trajectory the life of other species. The active principle here is not the curiosity of the human observer whose wanderings have led him to this scene of "wildness and variety," but rather the "will," the "taste," and the "joy" of nature as curator of her own pleasures, seemingly for her own sake—pleasures which in this instance centre on the expansive and untrammelled layering of foliage, a signal of depth not form.

In the lines that follow, Clare seizes upon other manifestations of "uncheck'd" (line 60) growth, other ornaments of wildness that cover Cowper Green in rich profusion, beginning not with its flowers but its weeds: "brambles," "thistle," "nettle," "hemlock," and "henbane" (lines 60–65), to be precise. It is a list practically drawn from the dark arts, suggesting an environment not merely indifferent but antithetical to human well-being. So palpable is this impression that the human reaction, as Clare points out, is typically one of "disgust" (line 69), at least from those inclined to "judge a blossom by the nose" (line 70). Human antipathy, however, also proves beneficial because the absence of "clowns" (and even other nature lovers like Clare himself) ensures the continued propagation of weeds, not only the noxious ones he names but "full many a nameless weed, / Neglected, left to run to seed" (lines 67–68). Embedded in this couplet is an indication—a trace—of the orderliness of wild nature, by which I mean not its compositional balance but rather its sense of purpose, or natural order, expressed in cycles of growth, decline, and reproduction. Letting things "run to seed" is the way of wildness, the way that "neglected" nature expresses itself and matures into a distinctive ecosystem according to rules, patterns, and protocols that appear to lie beyond the ordering framework of human aesthetic or instrumental understanding. As Clare suggests, only those with a "discerning eye" (line 99)—meaning those who through habit, patience, and humility have put themselves in a position to be retrained by wildness—can begin to perceive that "flowers of waste, / Planted here in nature's haste, / Display . . . / Her loved, wild variety" (lines 97–100). The modern term for this varied proliferation is biodiversity but Clare also inflects the idea with an aesthetic component through the word "display" which implies curation and perhaps even the anticipation of audience. Once again, beneath the veil of nature's disorderliness, its apparent "waste" and "haste," there is purpose and design.

While that purpose, as I have noted, is initially characterized as being inimical to human interests or at the very least indifferent to them, Clare also reveals that the layers of wildness, evident in the richness of its soil, contain traces of erstwhile human presence. The "uneven, heathy ground" (line 116) of Cowper Green periodically casts up "Antique coins of various kind" (line 126), mouldering "coffins in the rock" (line 128) and even "skull and bone" (line 129). Whereas human history's "early page," as Clare notes, too often "yield[s] to black oblivion's rage" (lines 145–146), the wild preserves within its depths a vast archive of relics and temporal strata that reach back to the druids and beyond. Indeed, the soil of Cowper Green, wild in itself, is also a history of wildness, a palimpsest on which the traces of human civilization are as readily evident as the cycles of growth and decline that characterize its flora. For Clare, a self-declared lover of "hoary, bald antiquity" (line 142), the coexistence of human and natural history speaks to a fundamental levelling of these narratives, these modes of being. They are not separate, irreconcilable, or mutually exclusive, as the picturesque engagement with wilderness implies. Human being in the wild is not a "problem" to be resolved by marginalization or excision; instead, humans, "nameless weed[s]," "flowers of waste," and countless other lifeforms coexist for Clare in a non-hierarchical relation of interdependence. The found fragments of history—these signals of the deep wild—in fact lead the poet on a journey of conjecture about the past that frames the relation between humans and nature as one of progressive decline. The "Hermits, fled from worldly care, / [Who] may have mossed a cottage" (lines 165–166) on Cowper Green in ages past, "Liv'd on herbs that there abound" (line 167) and these herbs, as Clare notes,

> . . . have existence still
> In every vale, on every hill,—
> Whose virtues only in them died,
> As rural life gave way to pride. (lines 169–172)

What wildness records in the layers of its own history is an order, now lost, that governed human being in the world of nature. Wildness was given freedom to "abound," while Clare's predecessors drew from it sustaining "virtues" in the form of "Food and physic" (line 168). He returns to this symbiotic relationship at poem's end and conjures up an image of the hermit having "thatch'd his roof, and made his bed" (line 188) from the same grasses and moss that still clothe the banks of Cowper Green. It is an image that in its temporal sweep anticipates the last lines of "The Flitting," Clare's great poem of displacement written after his relocation from Helpston to Northborough in 1832. That poem ends on a note of unexpected consolation as Clare finds solace in the same elements that feed his spirit on Cowper Green: the "Poor persecuted weeds [that] remain" and "the grass [that] eternal springs / Where castles stood and grandeur died" (lines 212, 215–16). In Northborough, as in the familiar spaces of Helpston, there is but one transcendent presence, one order that prevails over "castles" and "grandeur," the objects of conventional aesthetic attachment, one dominion that supersedes human life itself, absorbing its fragments and bringing up out of them the deep irrepressible wildness of unnamed weeds.

Works cited

Andrews, Malcolm. *The Search for the Picturesque: Landscape Aesthetics and Tourism in Britain, 1760–1800*. Stanford UP, 1989.

Barrell, John. *The Idea of Landscape and the Sense of Place, 1730–1840: An Approach to the Poetry of John Clare*. Cambridge UP, 1972.

Bate, Jonathan. *John Clare, A Biography*. Picador, 2003.

Bermingham, Ann. *Landscape and Ideology: The English Rustic Tradition, 1740–1860*. U of California P, 1986.

Berry, Wendell. "Life Is a Miracle." *Communio: International Catholic Review*, vol. 27, 2000, pp. 83–97.

Broglio, Ron. *Technologies of the Picturesque: British Art, Poetry, and Instruments, 1750–1830*. Bucknell UP, 2008.

Brook, Isis. "Wildness in the English Garden Tradition: A Reassessment of the Picturesque from Environmental Philosophy." *Ethics & the Environment*, vol. 13, no. 1, 2008, pp. 105–119.

Brownlow, Timothy. *John Clare and Picturesque Landscape*. Clarendon, 1983.
Carlson, Allen. "Aesthetic Appreciation of Nature and Environmentalism." *Royal Institute of Philosophy Supplement*, vol. 69, Oct. 2011, pp. 137–155.
Clare, John. "Cowper Green." *The Village Minstrel, and Other Poems*, vol. 1. Taylor and Hessey, 1821, pp. 108–120.
———. "The Flitting." *Poems of the Middle Period, 1822–1837.* Ed. Eric Robinson, David Powell and P.M.S. Dawson, vol. 3, Clarendon, 1989, pp. 479–489.
———. "Pilsgate Hill." *The Early Poems of John Clare, 1804–1822*, edited by Eric Robinson, David Powell, and Margaret Grainger, vol. 2. Clarendon, 1989, p. 118.
Copley, Stephen and Peter Garside. "Introduction." *The Politics of the Picturesque: Literature, Landscape and Aesthetics since 1770*, edited by Stephen Copley and Peter Garside. Cambridge UP, 1994, pp. 1–12.
Gilpin, William. *Observations on Several Parts of Great Britain, Particularly the Highlands of Scotland, Relative Chiefly to Picturesque Beauty, Made in the Year 1776*. 3rd ed., 2 vols. T. Cadwell and W. Davies, 1808.
———. *Observations on the River Wye, and Several Parts of South Wales, &c., Relative Chiefly to Picturesque Beauty, Made in the Summer of the Year 1770*. 2nd ed. R. Blamire, 1789.
———. *Three Essays: On Picturesque Beauty; On Picturesque Travel; and on Sketching Landscape: To Which Is Added a Poem, on Landscape Painting*. 2nd ed. R. Blamire, 1794.
Gorji, Mina. "John Clare and the Poetics of Mess." *Moveable Type*, vol. 5, 2009, pp. 1–11.
Hutchings, Kevin. "Ecocriticism in British Romantic Studies." *Literature Compass*, vol. 4, no. 1, 2007, pp. 172–202.
Irvine, John and Mina Gorji. "John Clare in the Anthropocene." *Cambridge Anthropology*, vol. 31, no. 1, 2013, pp. 119–132.
Khalip, Jacques. "Contretemps: Of Extinction and Romanticism." *Literature Compass*, vol. 13, no. 10, 2016, pp. 628–636.
Kövesi, Simon. *John Clare: Nature, Criticism, History*. Palgrave Macmillan, 2017.
Kutchen, Larry. "The 'Vulgar Thread of the Canvas': Revolution and the Picturesque in Ann Eliza Bleecker, Crevecoeur, and Charles Brockden Brown." *Early American Literature*, vol. 36, no. 3, 2001, pp. 395–425.
Price, Uvedale. *Essays on the Picturesque, as Compared with the Sublime and the Beautiful; and, on the Use of Studying Pictures for the Purpose of Improving Real Landscape*. J. Mawman, 1810.
White, Adam. "John Clare: 'The Man of Taste.'" *John Clare Society Journal*, vol. 28, 2009, pp. 38–54.
Wordsworth, William. *The Prelude: 1799, 1805, 1850*. W.W. Norton & Co., 1979.

4 Savage, holy, enchanted

Coleridge in concert with the wild

Gregory Leadbetter

"Amid the profoundest and most condensed constructions of hardest Thinking," Coleridge wrote in his notebook around 1820, "the playfulness of the Boy starts up, like a wild Fig-tree from monumental Marble" (*Notebooks* IV: 4777)—and indeed, the irrepressible, irruptive life of the wild abounds throughout Coleridge's writings. The word "wild" at once signifies a form of being—a dynamic state, whether inwards or outwards, beyond our deliberate control—and a relation to that state, capable of connoting fear and desire, subjection and freedom, threat and wonder. This essay pursues the implicate pattern in Coleridge's physical, intellectual, and imaginative relation to the wild—a self-directed exposure to "the influxes / Of shapes and sounds and shifting elements" that he wrote of in "The Nightingale" (*Poems* 161)—and the poetic principle, fundamental to every branch of his thinking, that emerges in that pattern. In doing so, it explores how the will can work in concert with the wild, as those lines from "The Nightingale" imply, and hence how those two ideas—of the wild and the will, commonly assumed to be diametrically opposed—can be brought into productive, mutually enlivening relationship. In poetry, this is to ask how order can act in concert with spontaneity—how art can transcend its own design—in the "interpenetration of passion and of will, of *spontaneous* impulse and of *voluntary* purpose" (*Biographia* II: 65), the dynamic union of the active and the passive, in which "measured Sounds" act as "the vehicle <& Involucrum> of Poetry" (*Lectures on Literature* I: 494), and the "*rules* of the IMAGINATION are themselves the very powers of growth and production" (*Biographia* II: 84).

While I draw upon writings from across Coleridge's career in developing this reading, I use "Kubla Khan" as a composite rune with which to organize my contentions—a poem where, as Coleridge wrote of Shakespeare's "Venus and Adonis," "You seem to be *told* nothing, but to see and hear everything" (*Biographia* II: 21). In its mythopoetic amplitude, "Kubla Khan" presents a microcosm through which to enter the macrocosm of the Coleridgean wild, and I take the poem as a psychotropic guide to the ways in which, throughout his work, Coleridge acknowledges, adapts to, and invokes the presence and the power of the wild as a living mystery implicate in human experience. The language of "Kubla Khan" also presents a relationship between the "savage," "holy," and "enchanted" that, as such, figures a relationship between the wild, the numinous, and poetic utterance—between the natural, the transnatural, and the power of language—that returns metaphysics to its poetic origin, in "the union of the sensuous and the philosophic mind" (*Shorter Works* II: 1267). The essay concludes by sketching the radical implications of this relationship and its reflex in Coleridge's later philosophical prose—at once a re-wilding and re-ordering at the "mysterious ROOT" of creative agency, that participates in "its darkness and its pregnancy" (*Lay Sermons* 50).

I turn first to the origin story of "Kubla Khan" itself: the famous preface with which it was published in 1816. Its subtitle nests the poem in "a vision in a dream," a "fragment" drawn from a wild source: a "profound sleep," "in which all the images rose up before him as *things*, with a parallel production of the correspondent expressions, without any sensation or consciousness of effort" (*Poems* 167, 163). Coleridge's dream-life has been the focus of much critical attention—not least because it was such a focus of his own (see Ford)—but my concern here is with what this example discloses with regard to the relation with the wild that I seek to describe. The older Coleridge lamented how little a lifetime's intense observation of dreams had yielded:

> O vanity! I have but a few hours back announced myself to my friend, as the author of a SYSTEM of Philosophy on Nature, History, Reason, Revelation; on the Eternal, and on the Generations of the Heaven and the Earth, and I am unable to solve the problem of my own Dreams! After many years' watchful notice of the phaenomena of the somnial state, and an elaborate classification of its *characteristic* distinctions, I remain incapable of explaining any one Figure of all the numberless Personages of this shadowy world—(*Letters* VI: 715)

That said, Coleridge never lost his sense that a "wild dream" might have "a deal of true psychological Feeling at the bottom of it" (*Notebooks* I: 1824). Alluding both to the Book of Job and to *Hamlet*, he writes in *The Statesman's Manual* that "'the visions of the night' speak to us of powers within us that are not dreamt of in their day-dream of philosophy": "a sort of under-consciousness blends with our dreams," where "our own self is the ventriloquist," while at other times, we do not dream "*about* things," but "*dream the things themselves*" (*Lay Sermons* 80, 81). The echo of the preface to "Kubla Khan," published the same year, is striking, and while it may superficially distance the will of the poet from the visionary experience and even the poem it describes—whatever the factual truth of the account—in its effect, the preface embodies Coleridge's abiding concern with acknowledging and establishing a dynamic *relationship* with the powers of spontaneous generation so intimately involved in the oneiric life of the mind. In the letter acknowledging that an explanation of the "shadowy world" of dreams eludes him, quoted above, Coleridge nevertheless notes that "A large part of the imagery of the Divina Comedia was in fact supplied by the Trances of a Monk, and it is certain that the monkish *ἄσκησις* may produce the magnetic sleep" (*Letters* VI: 715). This refers to the *cultivation* of "the magnetic sleep," a "trance" or altered state of consciousness, conducive to visionary insight: in other words, a process *enabled by the will*, that might turn the hypnagogic state into a psychotropic state, working in concert with the more or less latent energies of our being. Elsewhere, musing again on the paradox of a state at once active and passive in the "*Clair-voyance* of the Zoo-magnetists," Coleridge conceived of a state for which no word exists, a state not of "sleep," but "ἄνθυπνος," "the opposite to Sleep, or Counter-sleep" (*Marginalia* V: 726): a state, that is, of *anthypnosis*, as if *hypnotized awake*, to a higher or more subtle order of wakefulness. In situating the poem in such a particular relation to an oneiric origin, the preface to "Kubla Khan" prefigures such processes: its use of the words "composed" and "composition" with regard to the spontaneous givenness of the poem implies a certain order of agency, as does the description of "a profound sleep, at least of the external senses" (*Poems* 163)—which implies that some "inner sense" (*Table Talk* I: 53) might be awake and active. As I describe below, the exceptional technical finish of the poem in itself embodies precisely such an intimate relation between artful pattern and spontaneous life: the *forming* and the *wilding* coeval in imaginative experience.

Throughout his writings, Coleridge is his own wild. He described himself in his fifties as "a *Terrae Filius*" (Gillman 12), a child of the earth, and his self-portraits are everywhere infused with the wild and the desires it evokes:

> How oft pursuing fancies holy
> My moonlight path o'er flow'ring weeds I wound,
> Inspired, beyond the guess of folly
> By each rude shape, and wild unconquerable sound!
> ("France: An Ode," *Poems* 213)

He could both affirm and mock this in himself—as in the "love-lorn man" of "The Picture, or The Lover's Resolution," who makes his way "Through weeds and thorns, and matted underwood . . . with wild foot," and "Worships the spirit of unconscious life / In tree or wild-flower.—Gentle lunatic!" (*Poems* 283–284). In celebrating the "Orphic song" of Wordsworth's *Prelude*, Coleridge recalls "all which I had culled in wood-walks wild" ("To William Wordsworth," *Poems* 301), and this prefigures the "wild-wood fancy" of "Idoloclastes Satyrane," the thinly veiled version of himself that appears in "A Tombless Epitaph" in 1809, which translates and adapts a passage from Gabriello Chiabrera:

> not a hidden path, that to the shades
> Of the beloved Parnassian forest leads,
> Lurked undiscovered by him; not a rill

There issues from the fount of Hippocrene,
But he had traced it upward to its source,
Through open glade, dark glen, and secret dell,
Knew the gay wild flowers on its banks, and culled
Its med'cinable herbs. (*Poems* 315)

As Morton Paley remarks, this is "recognizably a version of the landscape of 'Kubla Khan,'" and describes a poet "to whom the natural world reveals its mysteries" (117). As in "Kubla Khan" (more implicitly), the emphasis on the "source" of poetry, traced through the inductive labyrinth of a living landscape, links the poet to the metaphysician: these are coeval in Coleridge's thinking. The pseudonym "Satyrane" is taken from Spenser—the "Satyres sonne yborne in forrest wyld" (*The Faerie Queene* I.v.21)—and Coleridge published the poem in *The Friend* as an introduction to "Satyrane's Letters," in which he describes his wanderings through Germany in 1798–99 (*Friend* II: 184–196). Coleridge at once observes, reflects upon, and projects himself in a suggestive fiction, which inflects another figure from his poetry: the foundling of "The Foster-Mother's Tale," discovered under a tree as a baby "wrapt in mosses," who combines an affinity for the wild with "heretical" learning and "unlawful thoughts of many things" (*Poems* 122).

Coleridge makes such a connection in relation to himself even earlier, in "The Eolian Harp," where the speaker of the poem basks on a hillside, indulging in "wild and various" thoughts that trouble the religious orthodoxy ventriloquized through Sara, his wife-to-be (*Poems* 53). Coleridge could be uneasy about the untrammelled life of his thoughts, too, but he nevertheless recognizes a vital nexus between his love of wild nature, "the embracement of rocks & hills," and the way "my spirit courses, drives, and eddies, like a Leaf in Autumn: a wild activity, of thoughts, imaginations, feelings, and impulses of motion, rises up from within me" (*Letters* II: 916). The phrase "wild activity" recalls the passage in "The Destiny of Nations" in which—based on his reading of the shamanic flight ("strange trance") of the Angekok, the "Greenland Wizard"—Coleridge directly relates the "wild activity" of the mind, the imaginative realization of forms that do not exist in nature, and the origins of spiritual growth:

For Fancy is the power
That first unsensualizes the dark mind,
Giving it new delights; and bids it swell
With wild activity; and peopling air,
By obscure fears of Beings invisible
Emancipates it from the grosser thrall
Of the present impulse, teaching Self-control
Till Superstition with unconscious hand
Seat Reason on her throne. (*Poems* 99)

This pattern goes to the root of Coleridge's abiding concerns, and what can be learned from his work. The "wild" and the ordering powers of the mind are not here diametrically opposed, but fundamentally related. "The End is in the Means" (*Letters* V: 98), and the play of the wild contains the principle of order with which it is coeval.

Coleridge's notebooks everywhere illustrate this principle in action. They are a wild realm within Coleridge's writings—a place where he was free to wander, a wild mind, "Happily disengaged & vacant never" as he once brilliantly put it (*Notebooks* I: 1697). The notebooks often reveal him physically in search of wild nature, to which his language bears an erotic relation, his attention led on by a fascinated experiential desire, as when walking through the "loneliness of a very narrow Valley" above Messina on 20 October 1805, in "deliciously aromatic" air, "more & more enamoured of the marvellous playfulness of the Surface of the Hills / such swellings, *startings*, sinkings" (*Notebooks* II: 2705). Again there are echoes of the ecstatic landscape of "Kubla Khan," the exhilaration of play and freedom, and the characteristic emphasis on the dynamics of origin—"*startings*"—as the "wild activity" of the living world meets that of the poet-philosopher, a mutual stimulus evoked and figured in his language. In his extraordinary attention both to the heterogeneous detail and the moods and relations of the natural world, Coleridge came up against the adequacy of language as a form of response to, and participation in, the complexity of its reality: "In Nature all things are individual; but a Word is but an arb[itrary Character] for a whole Class of Things; so that the same description may in almost all cas[es be applied] to twenty different appearances" (*Letters* I:503). The verbal superabundance of the notebooks, however—and indeed the whole range of his writings—can be read as a response to that problem, *desynonymizing* experience, as Coleridge says that language generates and refreshes itself like the division of cells in microscopic lifeforms (*Biographia* I: 82–83). Coleridge's delight in the self-ramifying processes of life is directly analogous to his delight in the self-ramifying processes of language and recognizes a fundamental relation between both.

This also connects Coleridge's notebooks with his poetry, as sites of mutual origination. One task of poetry, in and through formal qualities that draw and enliven the attention, is to invoke and embody in *language* an "individuality"—that is, a "wild life"—akin to the "individuality" in nature to which Coleridge refers. Just as (ideally) this process of poetic fascination awakens qualities of attention more alive to the differential qualities of reality, so in its very "yearnings & strivings" (*Notebooks* II: 2509) to think and to utter, to participate in what they see, the spontaneity of Coleridge's field-notes involves the same wild ground-work as his poems. Coleridge insisted that "Nature has her proper interest; & he will know what it is, who believes & feels, that every Thing has a Life of its own, & that we are all *one Life*" (*Letters* II: 864): an acknowledgement not only of relatedness but difference, of the intrinsic value of the *other*ness and unknown life both within and beyond ourselves. Life, inherently wild, has a value beyond utility, as an end in itself.

In his recognition, both in theory and practice, of the "individuality" of the natural world, Coleridge might agree with John Fowles that the "subtlest of our alienations" from nature, "the most difficult to comprehend, is our need to use it in some way, to derive some personal yield": we should, he writes, "dissociate the wild from the notion of usability" (39, 40). For Fowles, as for Coleridge, this involves the same kind of Kantian moral imperative that dissociates the idea of a person from the notion of usability: "a person can never become a thing, nor be treated as such without wrong . . . the latter may rightfully be used, altogether and merely, as a *means*; but the former must always be included in the *end*, and form part of the final cause" (*Friend* I: 190). This, in itself, is to cherish the wild—that is, as Fowles puts it, the "irreducible 'wild' component" of our being, which "can never be wholly plumbed, predicted or commanded" by science or society. Ordinary experience, Fowles writes, is "quintessentially 'wild'" in its complexity, and "corresponds very closely . . . with wild nature" (37). To venerate the "wild," the "individuality" in the natural and what Coleridge called the "Individuum" in the human (*Notebooks* V: 6487), as described here, is to venerate both a particular order of being and our relation to it: the productively complex, unknown, and even unknowable dimension both in our habitat and ourselves. To do so, moreover, is to begin to work in concert with the wild, the ineffable life in which we participate, and this involves an act of will, not as an attempt to force, fix, reduce, or delimit but to open an enabling, receptive, seminative, and mutually quickening relation: "will," that is, closer to the sense proposed by Iain McGilchrist, "a disposition towards the world" that characterizes our "primary being" (171). This discloses the challenge central to my theme: the task of *cultivating* our relation to the wild, and its bearing on the cultivation of ourselves, our habitat, and—with regard to poetry in particular—the powers and properties of language.

Coleridge is an explorer of the wild within and without the self, and—impatient with the "tacit compact among the learned, not to pass beyond a certain limit in speculative science" (*Biographia* I: 148)—he knew that this made his "hobby-horse," "metaphysics and psychology" (*Biographia* I: 85), a somewhat transgressive pursuit in the religious and intellectual context of his time. While this can give his public (and indeed private) voice a self-protective obliquity, his defences of metaphysics (in the sense that he used the term) blaze from his publications and his notebooks: those "minds that feel the riddle of the world" feel for the hidden grounds of experience, and in this the poet and the philosopher alike should "awaken the minds of others" to a "freshness of sensation" in relation to the familiar world (*Friend* I: n109–110). Indeed, Coleridge's poetics and metaphysics irradiate the known world with the unknown:

> I do not like that presumptuous Philosophy which in its rage of explanation allows no xyz, no symbol representative of the vast Terra Incognita

> of Knowledge [sic], for the Facts and Agencies of Mind and matter reserved for future Explorers / while the ultimate grounds of all must remain in-explorable or Man must cease to be progressive. Our ignorance with all the intermediates of obscurity is the *condition* of our ever-increasing Knowledge. (*Notebooks* III: 3825)

The unknown is essential to the possibility and valency of knowledge, and the animation and evolution of our becoming depend upon the *activation* of the unknown as a presence in our knowing: a task in which poetry and philosophy share. Far from an irresponsible indulgence in the irrational, this relation is essential to the idea of "reason" for Coleridge:

> In the most reflecting minds there may, nay must, exist a certain '*reserve* of Superstition['], from the consciousness of the vast disproportion of our knowledge to the terra incognita yet to be known . . . [this] constitutes the reason of *Superstition*, and makes it reasonable. (*Marginalia* IV: 579)

The mind must acknowledge and adapt to the reality of the "terra incognita"—the "unknown land"—within and beyond ourselves, to the affective presence of the spontaneous life in which we participate, the "wild" beyond our ken and control. Just as its "instinct . . . impels the chrysalis of the horned fly to leave room in its involucrum for antennæ yet to come" (*Biographia* I: 242), so in taking responsibility for its own development, the human mind should acknowledge and adapt not to "an other World, that is to come," but "the other world, that *now* is" (*Shorter Works* I: 569): the *esoteric* ecology, the "other world" *present* in our familiar habitat, however ostensibly ordered that habitat may be.

These "wild" bearings branch through "Kubla Khan," in the connections the poem makes between the "gardens" of the "pleasure-dome," the "savage," and the "romantic." Again, the union of the wild with the garden may seem superficially paradoxical, but the poem brings the two together. The "wilderness-plot, green & fountainous & unviolated by Man" (*Notebooks* I: 220) jotted in Coleridge's notebook, to which the poem so clearly relates, already contains, in that word "plot," the implications of an ordering, creative presence, correspondent with Sidney's description of "Poesy" as "an imaginative ground-plot of a profitable invention" (124). Both the wild and the garden share a secret life, in both natural and literary history, and the garden is a space for intimacy with the wild and intensity of encounter. In "This Lime-tree Bower my Prison," for example, the garden lime-tree bower accentuates and even hyper-develops the *inward* wild of the poet's imagination, which inductively returns him to his immediate location, with a heightened degree of attention: "No plot so narrow, be but Nature there, / No waste so vacant, but may well employ / Each faculty of sense" (*Poems* 120). Literary and contemplative space can likewise be its own *hortus inclusus*: Coleridge writes that "I found in my Books and my own meditations a sort of high-walled Garden" (*Letters* III: 216), another space to grow "a green thought in a green shade" (Marvell 158). "Paradise," the final, resonant word of "Kubla Khan," derives etymologically from the ancient Persian *pairidaeza*, an "enclosed park, orchard, or pleasure ground" (*OED*, 3rd ed.). The act of demarcating the garden-space might mark it as a *temenos*, a space made sacred by an ordering act in sympathetic relation to the life beyond itself—a demarcation of a microcosm that might aid the contemplative apprehension of the "world," the entirety of reality, the macrocosm. Coleridge considered "Gardening" as one of the fine arts coeval with and generative of "civilized life" (*Letters* III: 418)—arts based not on physical convenience or reductive notions of utility, but which "have their origin in the human Mind," in "ποιησις" ("making") and "containing the End in the Means" (*Lectures on Literature* II: 219; *Friend* I: 465n).

The wildness of "Kubla Khan," therefore, does not necessarily equate to wilderness, even if Coleridge and others often use the terms synonymously. The connotations of both words are, of course, culturally and contextually constructed. As Jay Griffiths observes, "wilderness" can be particularly loaded, especially in the context of colonial history: "In many parts of the world, indigenous people have lost their land for the idea of 'wilderness'"—as in the concept of *terra nullius* and its modern variants (266). If wilderness now tends to have exhilarating implications in Euro-American culture—as Christabel sleeps in Geraldine's arms, "Like a youthful hermitess, / Beauteous in a wilderness" (*Poems* 202)—it is due largely to the Romantic inversion of an older, more hostile attitude to the wild, often rooted in Judaeo-Christian texts. As Peter Larkin notes, it is possible "to recognize contemporary ecology as a uniquely romantic offspring" (77). What Mary Webb calls "the gospel of earth," where "we are swept up into the wild heart of the wild," is a cultural, and culturally mediated, response to the wild (128, 127). Ted Hughes writes that "Every new child is nature's chance to correct culture's error" (149)—but of course this in itself is a cultural idea and attitude. Thankfully, the crude dichotomy between "nature" and "culture"—to which, as a poet, Hughes does not, in practice, conform—is now increasingly challenged (see Haila). Certain kinds of human behaviour may destroy elements of the natural world, but as Griffiths says, it is also true that "culture protects nature" (45).

"Kubla Khan" embodies a cultural relation to the wild in the form of a poetic principle—a principle of *making*—which, echoing the rhythmic figurative drama in the fabric of the poem itself, suffuses Coleridge's later thinking. In *Aids to Reflection*, Coleridge writes that "A reflecting mind is not a flower that grows wild, or comes up of its own accord" (22)—but this, on the argument I am advancing, is not to *oppose* the wild: rather, it is to emphasize the vital importance of *cultivating* the inner powers in concert with the spontaneous energies of growth. It should be read together with this passage in *The Statesman's Manual*: "what the plant *is*, by an act not its own and unconsciously—*that* must thou *make* thyself to *become*!" (*Lay Sermons* 71). The process of cultivation—in this poetic character—became a central principle of Coleridge's spiritual and political philosophy: it is, he insisted, possible to be "perilously over-civilized, and most pitiably uncultivated!" (*Friend* I: 500). In "*cultivation*, in the harmonious development of those qualities and faculties that characterize our *humanity*" are the grounds of both the "permanency" and the "progressiveness" necessary to the well-being of society (*Church and State* 42–43). While their contexts differ, these patterns in Coleridge's writings anticipate aspects of the work of the Indian ecologist Vandana Shiva, who opposes the definition of the "wild as free of humans, and the cultivated as free of Nature": for Shiva, the wild is the "self-organisation of life," and in cultivating this "self-organising energy," we are "cultivating the wild." As Coleridge well knew, "culture" and "cultivation" share an etymological root in the Latin *cultus* and *colere*, whose meanings fuse the cultivation of land, ritual observance, and the education of the person. Coleridge emphasized the latter as the grounds of all human activity, which ultimately helped him to avoid sentimentalizing the countryside and country life: "A curious & more than curious Fact that when the country does not benefit," he went as far to reflect in 1803, "it depraves" (*Notebooks* I: 1553). This, in turn, fed into the controversy with Wordsworth over poetic diction (*Biographia* II: 42–45, 82), where Coleridge's point is that the cultivation of the mind and feelings works in concert with and patterns our response to the life beyond and indeed within ourselves, and in this it is animated by the will: an enabling act or disposition within ourselves that directs and orders, without controlling or determining, that relationship. To cultivate the wild in the poetic sense I have described—at once enacted and dramatized in the lyric myth of "Kubla Khan"—is to cultivate this form of self-directed becoming: to mingle and fuse the artful and the natural.

Our response to the "savage place"—and indeed the word "savage"—in "Kubla Khan" is inflected both culturally and contextually, in and through the poem. As with "wild" and "wilderness," "savage" has been used historically in a pejorative sense, but that sense is both recognized and rewritten by the poem. Coleridge often uses the word "savage" as synonymous with the "wild," as in his relish for "the wild betongued savage mountained" lakeland landscape of Cumbria in 1799 (*Notebooks* I: 511), and the "savage unforgettable Scene" he found in Sicily in 1804, which with its "savage women of the Torrent" and its "Paradise," combining a "great wall" with a wild landscape, again echoes "Kubla Khan" (*Notebooks* II: 2177). As with the "wild," the "savage" (derived from the Latin *silvaticus*, woodland) need not denote the absence of culture: culture has, in Coleridge, its "wild-wood" ways (*Poems* 315). When he writes that universally, as a species, whether or not "in savage states," "Man will not be a mere thing of Nature," but "will be & will shew himself a power of himself" (*Notebooks* III: 3339), in the power and impulse to modify both himself and his natural habitat, Coleridge does not oppose the human and the natural; rather, he identifies humankind as an essentially poetic being, "in a secondary sense *his own creator*" (*Lectures on Literature* I:192). *Poiesis—culture*—is transnatural: self-altering, creative and, like "mysterious Pan," capable of embodying an "intelligence blended with a darker power, deeper, mightier, and more universal than the conscious intellect of man" (*Biographia* II: 117).

"Kubla Khan" is a touchstone for how Coleridge's work reveals ways of acknowledging, adapting to, and acting in concert with the presence and the powers connoted in the "wild." The poem is inherently wild, in that its abiding fascination at once arouses imaginative and interpretative response, and defies readings that would reduce it to a paraphrasable meaning: indeed, it shares this second quality with *any* poem, insofar as it is a poem, and not merely a statement or a record of something outside itself. Coleridge kept the poem "wild"—part of his personal esoterica—for around eighteen years before publication in 1816. Presented as a "fragment," it is (akin with all fragments) a whole that gestures to what is or has become a "wild" origin, a displaced or altered entirety. The fragment draws attention to this latter quality, to which Coleridge was peculiarly attracted: "It is," he writes, "for the *fragmentary* reader only that I have any scruple" (*Aids to Reflection* 244–245). "Kubla Khan" is its own creation myth, and a creation myth of the characteristically Coleridgean fusion of poetics and metaphysics whose patterns I have developed as my theme.

"Kubla Khan" is composed as and grounded in ecstatic psychotropic experience, embodied not only in the fabric of the poem, but foregrounded in the preface, where the poet's reference to the use of an "anodyne"—opium—subtly connects the poet himself with the ecstatic poet-figure who "on honey-dew hath fed, / And drunk the milk of Paradise" (*Poems* 168), and this would seem to be affirmed by the poem's verbal resonance with Coleridge's letter to his brother George in March 1798: "Laudanum gave me repose, not sleep: but YOU, I believe, know how divine that respose [*sic*] is—what a spot of inchantment, a green spot of fountains, & flowers & trees, in the very heart of a waste of Sands!" (*Letters* I: 394). Coleridge was at once fascinated and unsettled by his own experiences of "second sight," orders of insight, being and becoming hidden in ourselves, whose epiphanies cross "the Boundary between the material & spiritual World" (*Notebooks* III: 4066, 3320). While "The Pains of Sleep" bears vivid witness to the "sufferings strange and wild" (*Poems* 291) in which his opium use involved him, Coleridge remained attuned to the ways in which opium could heighten his experience of transnatural vision—how, he writes in 1808, a "*pernicious Drug*" can make a person "capable of conceiving & bringing forth Thoughts, hidden in him before, which shall call forth the deepest feelings of his best, greatest, & sanest Contemporaries," as "proved to him by actual experience" (*Notebooks* III: 3320). Earlier in 1808, Coleridge sketches an idea for a poem in which the Fall of humankind is equated with the loss of "the Heavenly Bacchus," the "dreadful consequences of the interspersed vacancies left in his mind by the absence of Dionysus," and the ambivalent recovery of Dionysiac experience in "Οινος"—"wine"—the embodiment of the god (*Notebooks* III: 3263). As late as 1824, Coleridge surreptitiously notes a coded reference to opium (in Greek letters) that connects it with the oracular vapour of the Pythia at Delphi: "The Oracle-mist and intoxicating vapour—yet an Oracle & truly Mem. mem. ΩΠΜ" (*Notebooks* IV: 5111). To cultivate this order of experience, at once physically and imaginatively, is quintessentially to cultivate the wild, to work in concert with the life of the self beyond itself.

In "Kubla Khan," the poem becomes an organ of ecstatic experience. In Coleridgean poetics, this is not to abandon artfulness or order: on the contrary, as I have argued, it is to invoke the spontaneous, self-organizing energies in which we participate, within the animating activity of its making. The "synthetic and magical power" of the imagination is "put in action by the will and understanding, and retained under their irremissive, though gentle and unnoticed, controul"—enabling, in the poetic process, the interfusion of "a more than usual state of emotion, with more than usual order," blending "the natural and the artificial" (*Biographia* II: 16, 17). Coleridge observes that "*Metre itself* implies a passion, i.e. a state of excitement" (*Letters* II: 812), and this pattern is reciprocal: echoing Edward Young, he remarks that even in its "wildest" creations, poetry has "a logic of its own, as severe as that of science; and more difficult, because more subtle, more complex, and dependent on more, and more fugitive causes" (*Biographia* I: 9 and n).

Both figuratively and rhythmically, "Kubla Khan" creates the ecstatic space-time of a continuous present. Alph, the sacred river, flows through the cosmos of the poem in an infinite loop: from its hidden springs to its hidden seas in the "caverns measureless to man," its "mazy motion" abides without end, weaving the "mingled measure / From the fountain and the caves." The poem enacts a time-warp of endless self-iteration, with the abyssal, pulsing source of the Alph at its heart: the orgasmic landscape, "with ceaseless turmoil seething, / As if this earth in fast thick pants were breathing," contains the fountain that "at once and ever . . . flung up momently the sacred river." The chasm is forever haunted and erotically charged by the "woman wailing for her demon-lover," both of whom, grammatically, are there and not there, both actual and unactual in the moment of the scene: the chasm is *as* "holy and enchanted" *as ever* was so haunted. These figurative, syntactical effects act in concert with the rhythmical texture of the poem, and this points to another paradox that illustrates my theme. It is widely acknowledged that metrically, "Kubla Khan," as Reuven Tsur notes, is "one of the most regular poems in the English language" (5). Tsur argues that the "hypnotic-ecstatic" quality of the poem fuses a more-than-usual rhythmic regularity with a more-than-usual figurative, grammatical, and thematic variety: the one "binds" and the other "diffuses," as Jim Mays puts it (109), concurrently. As Tsur's pioneering readings describe, the poet's manipulation of metre and rhythm can generate, simultaneously, the productive amplification of both rational and supra-rational psychophysiological effects, fusing certainty and uncertainty, the willed and the unwilled, the "regular" and the "wild" (100–114). In this way, the continuous present of the poem created by the dance of its language generates a quality of trance: its supra-naturalistic, modulating patterns activate and heighten attention in participatory, in-the-moment experience. This is the "wildness" in the poem that cannot be paraphrased, for which there is no substitute beyond the poem itself: as with any of the fine arts, you have to be *within* its experiential matrix to know it.

In "Kubla Khan," the poetic fascination coeval with the continuous present of the poem is a form of enchantment. Poetry itself, for Coleridge, is "enchanted ground" (*Shorter Works* I: 649), and the poem embodies its own connection between the human, poetic activity of enchantment and its wild ground: the "holy" and "enchanted" are implicate in the "savage" nucleus of the poem, and its "romantic" character. To enchant and to charm derive from the Latin *cantare*, and (like the verb to which it relates) *carmen* connotes at once poem, song, and spell. "Kubla Khan" recombines these qualities, and enacts a hierophany achieved in its own making. The ostensibly conditional projection of this hierophany ("Could I revive within me . . .") belies the action of the poem. The poem's self-iterating structure as a continuous present contains both the realizing principle of its own self-altering state, and its realization: the "dome in air" *has* been "built" and envisioned in "music" in the fact of the poem—and, fascinated and altered in communion with the wild of his own transnatural vision, the poet-figure himself has become an ecstatic, daemonic being, "feared and shrunk from as a something transnatural" (*Notebooks* III: 4066).

Just as the touch of Geraldine's body magically alters Christabel psychophysiologically—"In the touch of this bosom there worketh a spell" (*Poems* 201)—and the Mariner's "strange power of speech" transforms his auditor (*Poems* 188), so the psychotropic potential of "Kubla Khan" touches the reader. The poem involves the reader in a gnostic relation with mystery: a knowing akin to wonder that acts as an agent of becoming. "In Wonder all Philosophy began," writes Coleridge, following Aristotle; "in Wonder it ends" (*Aids to Reflection* 236). In this sense, the poem acts as a wilding of the self, creating in the continuous present of its trance "*the true witching-time*" of the mind (*Friend* II: 117). "Romantic" poetry likewise, in the context I describe, is a form of re-wilding, in and through language. Addressing "the *excitability* of the human mind" (*Notebooks* III: 3827), it seeks to work afresh with the "wilderness of vocal reeds," the origins of language within itself, rather than language "mechanized as it were into a barrel-organ" (*Biographia* I: 38). *Forming* and *wilding* meet in one and the same moment, like the "wild Ducks *shaping* their rapid flight in forms always regular" that Coleridge used as an "image of Verse" (*Letters* II: 814), and the union, in poetry, "of *spontaneous* impulse and of *voluntary* purpose" (*Biographia* II: 65). The poem should act upon the reader as a form of life, with an animating touch. In this transnatural *poiesis*, the poet produces *another wild*: as Sidney writes, the poet generates "another nature" (100).

This is to resist the conventional opposition of the poetic and the natural, the cultural and the wild, still widespread among commentators. To animate language with qualities analogous to living things is a poetic, cultural achievement of a particular order. This point goes to the heart of the tension between humanism and ecocentrism, and my intention here—through the example of Coleridge—is to show the ways in which human activity can work in concert with the wild, quickening the life in which we participate: a dynamic continuity between the natural, the imaginal, and the artful.

This is also to oppose those readings of "Kubla Khan," and of poetry more generally, that persist in conceiving of poetic inspiration as something entirely alien from the agency of the poet. Poets are not entirely in control of the "wild" with which they work, but neither are they "*Automaton*-writers" (*Notebooks* IV: 5118). "Kubla Khan" does not simply replay Platonic ideas of poetic madness (see Dodds 80–82)—a sort of absolute "wildness"—but rather reconceives that tradition in a new relation, in which the will of the poet plays the elective, enabling role in relation to the wild that I have described. In this, as in so much else, Shakespeare is Coleridge's exemplary poet. Together with Schlegel, Coleridge argues against the commonplace eighteenth-century view of Shakespeare as *merely* a "wild, irregular, pure child of nature," and insists that his "Genius" worked in concert with his "Judgement" (*Lectures on Literature* I: 79, 142), but his point goes further. Coleridge's poetics and metaphysics condense and meet in vivid, concentrated form in his ideal of Shakespeare as "a Nature humanized, a genial Understanding directing self-consciously a power & a[n] <implicit> wisdom deeper than Consciousness" (*Lectures and Literature* I: 495).

The implications of those words contain the "true wild weird spirit" (Allsop I: 94–95) that I have pursued through Coleridge's writings, and signal its larger ramifications for poetry and philosophy. As an elective, motive power, the will is intricately involved in establishing a relationship to the unwilled, to the life beyond itself—at once capable of activating and conducting the spontaneous energies of the wild, as what Coleridge called (following Bacon) "*the leading Thought*" or "Initiative" (*The Friend* I: 455). This is the dynamic that lay, for Coleridge, at the heart of metaphysics: "how, being acted upon, we shall act; how, acting, we shall be acted upon" (*Letters* II: 949). Like the suggestive etymological connection between "will" and "wild" (Griffiths 49–50), such a conception of the will points to the coeval relationship of spontaneity and pattern. In this light, the will is more akin to a quality of attention: a cultivated receptivity whose creative agency acts as an animating infusion of attention. Its activity does not reduce its relation with the wild to the utility of physical convenience, but affirms its affinity with what Coleridge recognized as "*high moral utility*, —the utility of poetry and of painting, and of all that exalts and refines our nature" (qtd. in Perry 192).

This, in turn, implies a personal, political, and spiritual imperative of as fundamental importance today as ever. The political dimensions of the wild are inseparable from the intellectual, aesthetic, and poetic. To nurture the wild is to nurture the self-originating life both of the natural world and of the unique individuality of the person. To educate is to cultivate, in the productive relation of these terms that I have described, and this is to cultivate the wild. Coleridge everywhere identifies the process with the arousal of self-organizing powers, by analogy with the natural world: "Education is to man what the transmission of Instinct is to animals—entwines Thought with the living Substance, the nerves of sensation, the organ of soul, the muscles of motion, and this, finally, with the *Will*" (*Inquiring Spirit* 80). Education, for Coleridge, is not to be confused with the accumulation of information, valuable as that is: "what is a liberal Education? That which *draws* forth and trains up the germ of free-agency in the Individual" (*Letters* VI: 629). This form of education is not limited to literature, but may involve direct, ostensibly "idle" communion with wild living forms: "Never," he said in 1813, "imagine that a child is idle who is gazing on the stream, or lying upon the earth; the basis of all moral character may then be forming" (*Lectures on Literature* I: 586).

To cultivate the wild is to cultivate a social good at its pre-social source. The order of poetry to which "Kubla Khan" belongs grants access to—in the continuous present that its form creates—an ecstatic experience that exposes the reader to aspects of being beyond the socialized, transactional personality. Invoking, arousing, and quickening the wild energies of becoming, such poetry acts as an organ of psychophysiological growth. In the transformative moment—the *present* and *presence*—it creates, the irreducible, supra-verbal figure of the poem exposes the experience of order to its own origination: a point of "perpetual Genesis" (*Marginalia* V: 714). The poem acts to "excite the germinal power" (*Friend* I: 473), the self-generating vitality at the mysterious root of all knowing. In this way, poetry works with and upon the arcana of our being.

Works cited

Allsop, Thomas. *Letters, Conversations and Recollections of S.T. Coleridge*, 2 vols. London, 1836.

Coleridge, S.T. *Aids to Reflection*, edited by John Beer. Routledge and Princeton UP, 1993.

———. *Biographia Literaria*, edited by James Engell and W. Jackson Bate, 2 vols. Routledge and Princeton UP, 1983.

———. *The Collected Letters of Samuel Taylor Coleridge*, edited by Earl Leslie Griggs, 6 vols. Clarendon P, 1956–1971.

———. *The Friend*, edited by Barbara E. Rooke, 2 vols. Routledge and Princeton UP, 1969.

———. *Inquiring Spirit: A New Presentation of Coleridge from his Published and Unpublished Prose Writings*, edited by Kathleen Coburn. Pantheon, 1951.

———. *Lay Sermons*, edited by R.J. White. Routledge and Princeton UP, 1972.

———. *Lectures 1808–1819: On Literature*, edited by R.A. Foakes, 2 vols. Routledge and Princeton UP, 1987.

———. *Lectures on the History of Philosophy*, edited by J.R. de J. Jackson, 2 vols. Routledge and Princeton, 2000.

———. *Marginalia*, edited by George Whalley et al. 6 vols. Routledge and Princeton UP, 1980–2001.

———. *The Notebooks of Samuel Taylor Coleridge*. Ed. Kathleen Coburn et al. 5 double vols. Routledge and Princeton UP, 1957–2002. English numerals relate to entry number.

———. *On the Constitution of Church and State*, edited by John Colmer. Routledge and Princeton UP, 1976.

———. *Poems*, edited by John Beer. Everyman, 1986.

———. *Shorter Works and Fragments*, edited by H.J. Jackson and J.R. de J. Jackson. 2 vols. Routledge and Princeton UP, 1995.

———. *Table Talk*, edited by Carl Woodring. 2 vols. Routledge and Princeton UP, 1990.

Dodds, E.R. *The Greeks and the Irrational*. U California P, 1951.

Ford, Jennifer. *Coleridge on Dreaming: Romanticism, Dreams and the Medical Imagination*. Cambridge UP, 1998.

Fowles, John. *The Tree*. Ecco P, 1983.

Gillman, James. *The Life of Samuel Taylor Coleridge*. London, 1838.

Griffiths, Jay. *Wild: An Elemental Journey*. Penguin, 2006.

Haila, Yrjö. "Beyond the Nature-Culture Dualism." *Biology & Philosophy*, vol. 15, 2000, pp. 155–175.

Hughes, Ted. *Winter Pollen: Occasional Prose*. Faber and Faber, 1994.

Larkin, Peter. *Wordsworth and Coleridge: Promising Losses*. Palgrave Macmillan, 2012.

Marvell, Andrew. *The Poems of Andrew Marvell*, edited by Nigel Smith. Routledge, 2013.

Mays, J.C.C. *Coleridge's Experimental Poetics*. Palgrave Macmillan, 2013.

McGilchrist, Iain. *The Master and His Emissary: The Divided Brain and the Making of the Western World*. Yale UP, 2012.

Paley, Morton D. *Coleridge's Later Poetry*. Oxford UP, 1996.

Perry, Seamus, ed. *S.T. Coleridge: Interviews and Recollections*. Palgrave, 2000.

Schlegel, A.W. *Vorlesungen über Dramatische Kunst und Litteratur*. Heidelberg, 1809–1811.

Shiva, Vandana. "Trusting and Cultivating the Wild." *Resurgence*, vol. 304, September/October 2017, www.resurgence.org/magazine/article4935-trusting-and-cultivating-the-wild.html

Sidney, Philip. *An Apology for Poetry*, edited by Geoffrey Shepherd. Manchester UP, 1973.

Tsur, Reuven. *"Kubla Khan": Poetic Structure, Hypnotic Quality and Cognitive Style*. John Benjamins, 2006.

Webb, Mary. *Poems and the Spring of Joy*. Jonathan Cape, 1928.

5 Human grapes in the wine-presses

Vegetable life and the violence of cultivation in Blake's *Milton*

Tristanne Connolly

Introduction: "infinitesimal grapes"

The first time William and Catherine Blake had their own garden, in Lambeth in the 1790s, they planted a grapevine (Bentley 745). According to visitor Maria Denman, "Blake would on no account prune this vine, having a theory it was wrong and unnatural to prune vines: and the affranchised tree consequently bore a luxuriant crop of leaves, and plenty of infinitesimal grapes which never ripened" (Bentley 745n). In this anecdote, Blake is so wild that he makes it a principle of liberty not to cut back the vine's growth; he cares more that it be "affranchised" than productive. But would Blake, or the vine, really be satisfied with stunted "infinitesimal grapes" as the fruits of freedom? Frederick Tatham's non-eyewitness report that when the Flaxmans drank tea in Blake's Lambeth garden "under the shadow of a Grape Vine which Mrs. Blake had very carefully trained . . . they all sat with ripe fruit hanging in rich clusters around their heads" (qtd. in Bentley 675) represents not wild but luxuriant Blake, which may be closer to the truth. After all, according to Milton, Adam and Eve would "lop" the "wanton growth" of the plants in the garden of Eden (IV:629), and the Blakes' garden in Lambeth is the one where they legendarily played Adam and Eve, sitting naked reciting *Paradise Lost* together (see Bentley xxvi-xxvii). If the truths of his Lambeth garden are unrecoverable, Blake's works give ample evidence of a much more complex "theory" on the tension between wildness and cultivation than Denman's anecdote suggests.

In *America a Prophecy* (1793), a product of Blake's Lambeth years, the "fires of Orc" have a remarkable effect on "the female spirits of the dead" who "Run from their fetters reddening" and "feel the nerves of youth renew, and desires of ancient times, / Over their pale limbs as a vine when the tender grape appears" (15:20–26).[1] Vividly, the liberty of the grapevine is in the growth of the fruit, which embodies sexual arousal, and a revolutionary kind of fertility. Orc is a "Lover of wild rebellion" (7:6), but he could not achieve such a yield without cultivation any more than Blake could in his garden. In the chapter "On Vines" in John Kennedy's *A Treatise upon Planting, Gardening, and the Management of the Hot-House* (1777), one finds phrases such as the following: "it will be of great service to add a little fire to them for two months after they are planted, as it would encourage their growth greatly" (1:226). Kennedy gives advice for grapevines grown on common walls, fire-walls (with a flue), and in hothouses warmed by fire; in England, grapes will only "ripen without fire" in the south (1:245). With or without fire, the vines need pruning for optimal growth. Kennedy argues against idealizing wild growing conditions. He writes, "I have been informed that the finest of the Spanish wines are produced from the grapes that grow wild on the rocks without any cultivation," but contends that "the grapes in those countries are small," and though "the great heat brings them to perfection" (a temperature impossible to replicate in England by fires), "they are not allowed to grow above three or four feet high, and to carry a few clusters on each branch," while "the wood, foliage, and fruit must be supported to bring them to perfection" (1:228–229). Kennedy's instructions echo Blake's Christlike mythology of Orc: "keep the young shoots nailed as they grow" (1:234), as Los nails down the boy Orc and binds him with the chain of jealousy (*Four Zoas* V, 60:28). The violence of cultivation is clear in Kennedy's vocabulary: "As soon as the Vines are pruned, let them be nailed" (1:238); old wood "will push out many shoots at every amputation . . . which must never be suffered to grow" (1:239); "very soon after the shoot is pulled off it begins to bleed" (1:237). Further, to stop the bleeding, the holes left by pulling off shoots "should be immediately filled up with worked clay" (1:236), recalling the creation of Adam and, in turn, Christ as the new Adam taking on a mortal body. If left to bleed, "it will hurt all the fruit on that branch, and greatly weaken the Vine" (1:236): the clay works as a mercy and a kind of salvation. In *Milton a Poem*, Blake has Milton take clay, "moulding it with care / Between his palms: and filling up the furrows of many years," to mercifully build a body for Urizen, "on the bones / Creating new flesh on the Demon cold, and building him, / As with new clay a Human form" (19:10–14).

Milton had its beginnings in the Blakes' three years in rural Felpham (1800–1803), where they moved at the invitation of Blake's patron William Hayley.[2] This was the only time they lived outside the metropolis. *Milton*'s agricultural concerns are established from the start in the Bard's Song which catalyses the subsequent action, and at the end, the pre-apocalyptic arrival of Milton and his emanation Ololon on earth takes place in Blake's Felpham cottage garden. Both of *Milton*'s books move towards "the Great Harvest & Vintage of the Nations" (43:1). *Milton* can be read as Blake's meditation on the garden and the wild, on the relation of "this earth of vegetation on which now I write" (14:41) to Eternity, and what kind of cultivation is necessary not only for grapevines and other plants, but also for humans, and all creation, to come to fruition. Since the Bard's Song deals more with the cultivators than the cultivated, here, after an initial exploration of cultivation in Blake, I would like to juxtapose two crucial passages in *Milton* on vegetable and human cultivation: the flower chorus, and the description of the

Vintage. The flower chorus shows how a blend of cultivation and wildness reveals the visionary eternal forms of plants in which they are endowed with human-like capacities of sensibility and creativity. The Vintage of the Nations, conversely, treats humans as plants, and shows how cultivation itself can go "wild" when its violence edges over the limit of salutary pruning into deadly harm. But the Vintage has the potential to reveal the eternal forms of humans, since it is the cyclical culmination of the perpetual planting of humans as seeds, and an apocalyptic event leading to the renovation of the earth. Because Blake, especially in *Milton*, plays freely with time, the Vintage is simultaneously a future cataclysm and an aspect of "vegetative" life on earth as a condition of suffering. In *Milton*, Blake purposely leaves unresolved whether the violence of the Vintage leads to the transformation of the human grapes or to tragic disaster; he avoids justifying violence by a redeeming result, and recognizes the contingency of all work of cultivation.

"Cultivated life"

T. S. Eliot memorably describes Blake as "a wild pet for the supercultivated," then dismisses the tag as reductive (151). Blake's wildness is indeed too easily assumed, but the paradox of a wildness tied to cultivation is suggestive. The word "cultivated" appears most often in Blake's exchange with Joshua Reynolds, where it concerns the cultivation of taste and talent in humans, and art in human societies. Blake contends, "The Bible says That Cultivated Life. Existed First—Uncultivated Life. comes afterwards from Satans Hirelings" (637). Counterintuitively, for Blake, the original state is not wild in the sense of living or growing in a state of nature (*OED* A.I); cultivation is not introduced by human effort to a pre-existing wilderness. But if the first humans were placed in the Garden of Eden to tend it, then "Uncultivated Life" equates to the Fall. Blake even writes, though he deletes it, "First were Created Wine & Happiness," connecting the grapevine to original cultivation, and insisting on pleasure. Blake is arguing against Reynolds's claim that cultivated life begins with necessaries and progresses to ornaments. Blake insists, "Satan took away Ornament first . . . Then he became Lord & Master of Necessaries" (637). Blake as an artist considers ornament essential, not an afterthought or surface addition, and, under Milton's influence, associates "Necessaries" with Satan's tyranny ("necessitie, / The Tyrants plea" [IV:394]). It seems that the Fall is the separation of Nature from Art and Imagination. For Blake this is not far from the more orthodox understanding that the Fall is the separation of humans and nature from the divine. In *A Vision of the Last Judgment*, Blake calls "the Saviour the True Vine of Eternity the Human Imagination" (555).

At the beginning of Reynolds's Discourse VII, Blake writes, "The purpose of the following Discourse is to Prove That Taste & Genius are not of Heavenly Origin & that all who have Supposed that they Are so. Are to be Considerd as Weak headed Fanatics," or, wild in the sense of not having control of one's mental faculties (*OED* 12a). Also proved is "that the Stupid are born with Faculties Equal to other Men Only they have not Cultivated them because they thought it not worth the trouble" (658). Blake disdainfully corrects Reynolds's misunderstanding of cultivation: it cannot develop anything that is not already there. "The Stupid" could not develop "Taste & Genius" by effort and training any more than a grapevine could become a clematis by persistent gardening. Commenting on the previous discourse, Blake brings together aesthetic and horticultural cultivation: "Reynolds Thinks that Man Learns all that he Knows I say on the Contrary That Man Brings All that he has or Can have Into the World with him. Man is Born Like a Garden ready Planted & Sown This World is too poor to produce one Seed" (E656). Gardens and cultivation, seeds and plants, for Blake have everything to do with identity. The kind of control and restraint wild Blake will not submit to is forcing a living thing against the unfolding and expression of its identity. Cultivation, properly understood, develops that identity and connection to that otherworldly source of all seeds. The implication is that cultivation of plants and humans is a similar process.

The word "cultivation" appears only once in all of the Illuminated Books, in *Milton*. "Allamanda calld on Earth Commerce, is the Cultivated land / Around the City of Golgonooza in the Forests of Entuthon" (27:42–43). Golgonooza, "namd Art & Manufacture by mortal men" (24:49), is surrounded by Commerce, which Blake considers "far from being beneficial to Arts" because it "Cannot endure Individual Merit its insatiable Maw must be fed by What all can do Equally well . . . as I have found to my Cost these Forty Years" (*Public Address* 573–574). Commerce is negative for shutting Blake out but letting inferiors in. But the *Milton* passage goes on to explain that "every Occupation of Men" is derived from the eternal arts of Poetry, Music, and Painting, otherwise "shut out" from this world (27:57, 60–62). The occupations, involved with commerce, allow the eternal arts to "Become apparent in Time & Space . . . That Man may live upon Earth till the time of his awaking" (27:57, 59, 61). Allamanda is a buffer zone around Golgonooza with the dual potential to restrict or allow access. Commerce, like cultivation, has power to determine what will grow and what will not. In the Annotations to Reynolds, Blake calls for a Commerce that would deploy this power properly. Reynolds gives "advice to those who are contending for royal liberality" of patronage, but Blake responds, "We want not Liberality We want a Fair Price & Proportionate Value," and he adds, "a General Demand for Art" (637). In Allamanda the cultivated land, "the Sons of Los labour against Death Eternal," and "Urizens sons here labour also . . . Here are the stars created & the seeds of all things planted / And here the Sun & Moon recieve their fixed destinations" (27:44, 49, 53–54). The growth of seeds on earth depends on the regular cyclical movement of "the Sun & Moon." Commerce and "fixed" order, both usually negative for Blake for their violence to individual identity and perception, here give form to what might otherwise, in the material world, be inaccessible or in danger of falling into formless chaos or even Eternal Death.

On the previous plate, Blake describes embodiment as form-giving cultivation:

> Souls incessant wail, being piteous Passions & Desires
> With neither lineament nor form but like to watry clouds
> . . . .
>
> The Sons of Los clothe them & feed & provide houses & fields
>
> And every Generated Body in its inward form,
> Is a garden of delight & a building of magnificence,
> Built by the Sons of Los in Bowlahoola & Allamanda. (26:26–33)

Again the human body itself is a garden; and the rescue from formlessness includes the provision of "fields" for cultivation. The garden for Blake is an emblem of nurturing limitation. In his art theory, Blake's resounding emphasis on the bounding line connects form-giving with identity—vegetable and animal as well as human—and with the original divine act of creation.

> How do we distinguish the oak from the beech, the horse from the ox, but by the bounding outline? How do we distinguish one face or countenance from another, but by the bounding line and its infinite inflexions and movements? What is it that builds a house and plants a garden, but the definite and determinate? . . . Leave out this l[i]ne and you leave out life itself; all is chaos again, and the line of the almighty must be drawn out upon it before man or beast can exist. (*Descriptive Catalogue*, 550)

When the Blakes planted their Lambeth garden, they had to impose the "definite and determinate" on space and vegetation that would otherwise revert to "chaos." If their grapevine had not been "carefully trained" but left to grow wild, it may have been free of constraint, but would not become fully itself; it would be stunted and disproportioned, with tiny grapes among too many leaves. The gardener, like the artist, and God as creator, perceives the form of each thing and helps it to emerge. The "bounding line" at once binds, with the painful subjection to control and restriction that implies, and is bounding, exuberantly springing with distinctive life.

The artist's line and the gardener's seed appear together in *Milton* in the description of the way one of the Sons of Los gives form to formless souls:

> The little weeping Spectre stands on the threshold of Death
> Eternal; and sometimes two Spectres like lamps quivering
> And often malignant they combat (heart-breaking sorrowful & piteous)
> Antamon takes them into his beautiful flexible hands,

As the Sower takes the seed, or as the Artist his clay
Or fine wax, to mould artful a model for golden ornaments.
The soft hands of Antamon draw the indelible line:
Form immortal with golden pen; such as the Spectre admiring
Puts on the sweet form. (28:9–18)

Gardening and art and human embodiment are interrelated processes of productive limitation. The Spectres are like seed, since Antamon takes them into his hands as a sower does. As the seed of a living thing is put into "clay" or soil to grow, similarly the artist makes a model in a perishable and less valuable medium, clay that is broken or wax that is melted, as a mould for the more lasting and precious: "golden ornaments" (remembering that "Ornament" is what Satan took away from original, cultivated, Edenic existence). Antamon's "indelible line" is the double-edged bounding line that limits and constrains yet gives shape and infuses Art and Imagination into Nature; it allows the Spectre to see the beauty of defined form and voluntarily take it on. "Form immortal" suggests that the form itself holds immortality. Blake writes, in *A Vision of the Last Judgment*, "the Oak dies as well as the Lettuce but Its Eternal Image & Individuality never dies. but renews by its seed. just so the Imaginative Image returns by the seed of Contemplative Thought" (555).

The flower chorus

Near the beginning of Book the Second of *Milton* appears this visionary scene of floral life:

First eer the morning breaks joy opens in the flowery bosoms
Joy even to tears, which the Sun rising dries; first the Wild Thyme
And Meadow-sweet downy & soft waving among the reeds.
Light springing on the air lead the sweet Dance: they wake
The Honeysuckle sleeping on the Oak: the flaunting beauty
Revels along upon the wind; the White-thorn lovely May
Opens her many lovely eyes: listening the Rose still sleeps
None dare to wake her. soon she bursts her crimson curtaind bed
And comes forth in the majesty of beauty; every Flower:
The Pink, the Jessamine, the Wall-flower, the Carnation
The Jonquil, the mild Lilly opes her heavens! every Tree,
And Flower & Herb soon fill the air with an innumerable Dance
Yet all in order sweet & lovely, Men are sick with Love! (31:46–62)

These flowers are dancers and musicians. Their synaesthetic music is scent which they orchestrate to "fill the air . . . all in order sweet & lovely." Though some are wildflowers, all are cultivated flowers in the sense of having developed their aesthetic taste and talent (and all have more innately to develop than "the Stupid" among humans). In response to the morning, these flowers have intense "Joy even to tears" that "opens in the flowery bosoms": Blake endows them with a seat of emotion. Further, he has them take part not merely in the rain cycle but in an ecosystem of sensibility: the Sun expresses pity when it dries the flowers' tears that were caused by the approach of dawn. This poignant joy seems to cue the leaders of the dance. Joy for Blake is intimately connected with the free expression of identity, as Oothoon most vividly shows in her ponderings in *Visions of the Daughters of Albion* on animals' diverse "habitations" and "pursuits, as different as their forms and as their joys," and her assertion, "are not different joys / Holy, eternal, infinite!" (3:5–6, 5:4–5). Each flower is specifically named, and the first several are given descriptions, as though each has a personality it is expressing through its odour and movement. All of these unique contributions build to a collective dance, overwhelmingly "Innumerable . . . / Yet all in order": the diverse individuals harmonize and only then are generally called "Tree, / And Flower & Herb." "Men are sick with Love!" parallels "Joy even to tears," suggesting that humans and plants can equally experience and mutually inspire bittersweet, complex emotion.

A few plates later, "The Wild Thyme is Los's Messenger to Eden, a mighty Demon / Terrible deadly & poisonous his presence in Ulro dark / Therefore he appears only a small Root creeping in grass" (35:54–56). Blake depicts a diversity of flower personalities, giving the "Terrible deadly & poisonous" a part in the flower chorus, and in the labours of Los. As "Messenger to Eden" the Wild Thyme is connected to the original state of cultivation. "Wild" is not so much the condition or location of the thyme as its name, the common name for *Thymus serpyllum*, while garden thyme is the common name for *Thymus vulgaris*, according to Miller's classic *Gardeners Dictionary* (1768); *Thymus serpillum* "grows naturally upon dry commons and pastures in most parts of England, so is very rarely admitted into gardens." *Thymus vulgaris* is "cultivated in the gardens for the kitchen" and "also grows naturally on stony rocky places in the south of France, in Spain and Italy." Garden versus wild is nominal for these plants. The flower chorus includes plants that may be found on uncultivated land, such as with the Wild Thyme and its companion the Meadow-sweet, but also ones more likely to be found in gardens, such as Jessamine, Carnation, and Lilly. Kevin Hutchings assumes the Wild Thyme's "presence in Blake's garden" so that this "*domestic* garden . . . this enclosed garden . . . incorporates *within* its domesticity something of the chaotic 'wildness' of non-domesticated nature" (150, 151). But where is non-domesticated wildness, especially in England where the landscape is visibly marked by millennia of human agriculture? Where is the uncultivated, for Blake, except in Satan's necessity, in nature removed from Art, Imagination, and Eternity? In the illustration on plate 36 of *Milton* (Figure 5.1), Blake's cottage garden does not appear to be enclosed; there is no wall or fence around it, and there is a clear view beyond it, through the trees on the right, to the sea. It is very much open above and beyond as well, with more

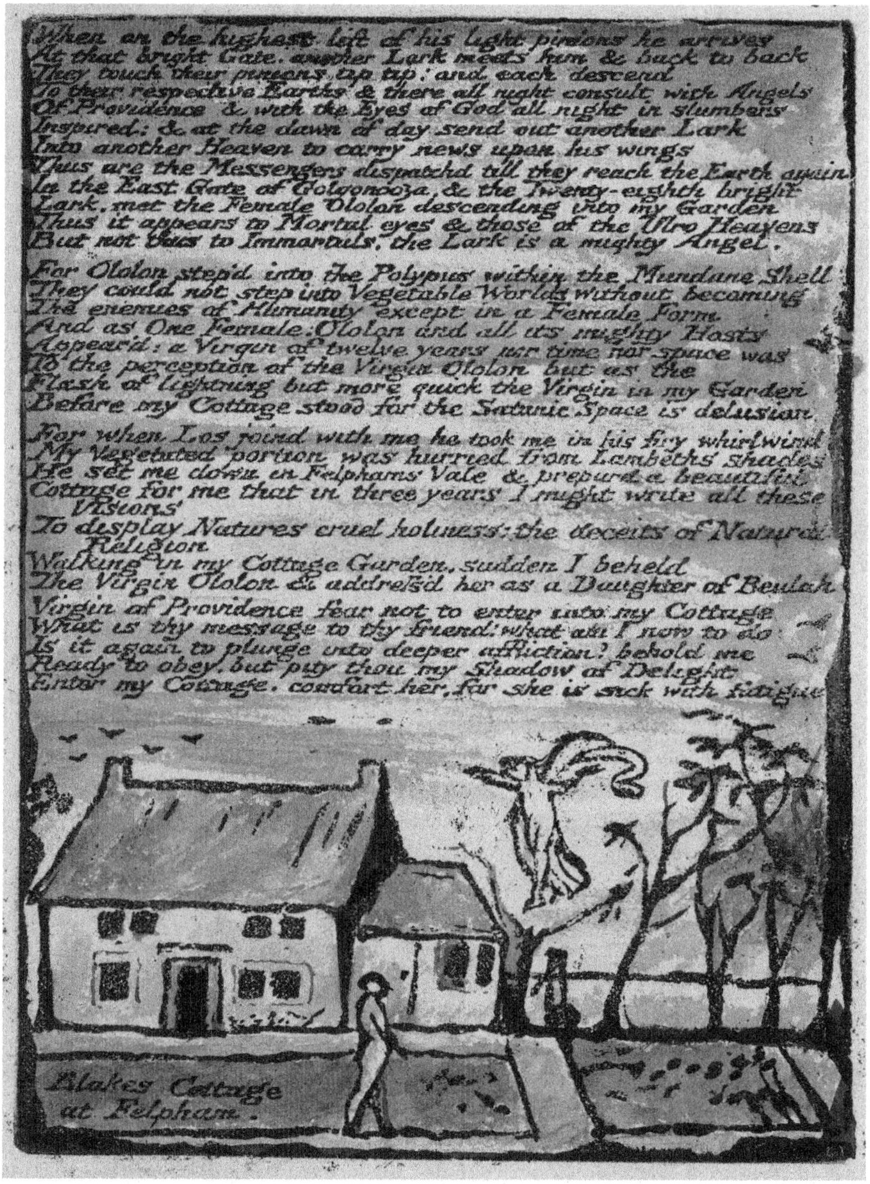

Figure 5.1 Plate 36, Copy B (ca. 1811), *Milton a Poem* by William Blake, 54041, The Huntington Library, San Marino, California (Blake Archive).

than half of the plate devoted to sky (and words), and with Ololon pictured descending into the garden from Eternity. On plate 35, Blake suggests the flower chorus does not exactly happen in his garden, or even in a place, but in a time that is somehow a place: “Just in this Moment when the morning odours rise abroad / And first from the Wild Thyme, stands a Fountain in a rock / Of crystal flowing into two Streams” which each run through various locales Edenic and Satanic, ultimately “Meeting again in Golgonooza” (35:48–51, 53). This same moment is also “a Moment in each Day that Satan cannot find / Nor can his Watch Fiends find it, but the Industrious find / This Moment & it multiply. & when it once is found / It renovates every Moment of the Day” (35:42–45), freeing “the Industrious” from measured “Watch” time to do their creative work. This moment is “Wild Time” for cultivation. If time is not mechanical but organic like thyme, then it can grow and spread; “multiply” suggests that it can be propagated like a plant.

The Vintage

Blake almost never uses “vegetable” or “vegetative” to refer to plant life,[3] but rather to the state of Generation, the earthly, material world, and the “narrow doleful form” of humans in Ulro: “Can such closed Nostrils feel a joy? or tell of autumn fruits / When grapes & figs burst their covering to the joyful air / Can such a Tongue boast of the living waters? or take in / Ought but the Vegetable Ratio & loathe the faint delight” (*Milton* 5:19, 32–35). Here, as in the flower chorus, ironically the “real vegetables” are bursting with joy, while the humans are more “Vegetable” in the stark limitation of their fallen senses (cf. Hutchings 117). Is vegetation a dire fate for humans, or a form-giving rescue, or a revelation of our continuity with other forms of life? The Vintage of the Nations in *Milton* intertwines all of these possibilities. It is described in detail in the final plates of Book the First, where the labours against Eternal Death discussed above also appear.

Blake takes the surreal image of human grapes being crushed for wine from Revelation 14:14–20, which elaborates on Isaiah 63:1–6. In both, the Vintage is a very bloody expression of divine wrath. Revelation adds the idea that “the time is come . . . to reap . . . gather the clusters of the vine of the earth; for her grapes are fully ripe,” as though the earth is one big vineyard planted and tended for this purpose. But it makes little sense to see harvesting grapes as an expression of the vintner’s anger at a mature and fruitful crop. On the contrary, the vintage is the consummation of all the labour and expectation of cultivation. It is terrifying for those being reaped and crushed, but that is how apocalypse as well as agriculture works; death and destruction precede new creation, and that is what vegetable matter is for, to grow, bear fruit, be plucked and cut down, and grow again next season. The Biblical passages, in likening vegetable to human life, admit the possibility that vegetable life suffers when it is harvested. Latent is a question Blake brings into relief: what if humans were treated by a higher being the way humans often treat animals and vegetables, to be grown for what they can yield? If human and plant life are analogous, and if plants too have imaginative eternal forms, then humans cannot expect their vegetative bodies to be uniquely exempt from the uses of nature and the suffering they involve.

Blake's boldest departure from Isaiah and Revelation is that he has animals and plants crushing the human grapes in the Vintage. I have argued elsewhere that the Vintage could be what Ron Broglio calls a scene from the animal revolution (Bruder and Connolly 15); it is equally a plant revolution to overthrow human oppressors. "[T]he little Seed; / The sportive Root . . . Dance round the Wine-presses . . . they rejoice with loud jubilee" (27:11–13, 23), and then after a catalogue of animals and insects, largely unpleasant, there is a focus on unpleasant plants:

> There is the Nettle that stings with soft down; and there
> The indignant Thistle: whose bitterness is bred in his milk:
> Who feeds on contempt of his neighbour: there all the idle Weeds
> That creep around the obscure places, shew their various limbs.
> Naked in all their beauty dancing round the Wine-presses. (27:25–29)

As with the flower chorus, each has its identity emphasized through description that ascribes emotion and volition to plants. "Naked" is a human quality in that only humans wear clothing, often made from vegetable matter such as flax. How can plants be naked? They may be throwing off their earthly coverings and displaying themselves in their eternal forms, no longer limited and subordinate. Their rejoicing is depicted as orgiastic revelling in violence, delighting in human suffering as humans would in theirs at a vintage of "vegetable" grapes. They are "drunk with wine" (27:24), again a human quality difficult to imagine in plants, especially with the further reversal that the wine is human blood; but, blood is a very good fertilizer. When the Nettle, Thistle, and Weeds are revealed "Naked in all their beauty," their opposition to humans is not dispelled but apocalyptically intensified. One definition of a weed is a plant "regarded as a nuisance in the place where it is growing, esp. when hindering the growth of crops or other cultivated plants" (*OED* I.1.a.a). But here the wild plants are doing agricultural work, a paradox that makes them akin to the cultivated Wild Thyme.

The contrast between the rejoicing non-humans and the suffering humans is intense: "But in the Wine-presses the Human grapes sing not, nor dance / They howl & writhe in shoals of torment; in fierce flames consuming" (27:30–31). But even though the human grapes are the only ones in "torment" in the Vintage, they are not the only ones dying.

> How red the sons & daughters of Luvah! here they tread the grapes.
> Laughing & shouting drunk with odours many fall oerwearied
> Drownd in the wine is many a youth & maiden: those around
> Lay them on skins of Tygers & of the spotted Leopard & the Wild Ass
> Till they revive, or bury them in cool grots, making lamentation. (27:3–7)

The Vintage is dangerous for the labourers too, but for the sons and daughters of Luvah, death—gruesomely drowning in blood-wine—is part of the process, or even the festivities. They soon "revive" after rest on skins of dead animals which seem to retain and communicate life force. It seems unlikely that even those buried in apparently refreshingly "cool grots" remain dead. The lamentation seems to be more of a formal response than an expression of inconsolable grief. In fact, the flower chorus, not a sad song at all, is presented as "a Vision of the lamentation of Beulah over Ololon," and is about seeing beyond "Mortal & Perishing Nature" (Annotations to Reynolds, 660). It could be that the plants and other creatures rejoicing in the Vintage are in danger of death too, but not distressed about it, such as "the Grasshopper that sings & laughs & drinks: / Winter comes, he folds his slender bones without a murmur" (27:19–20). Not only are they more used to considering themselves subject to natural cycles than humans tend to be; also, they have less to lose as their earthly forms are so lowly. If humans consider themselves exempt and in charge in relation to the rest of nature, they could easily come to think of the destruction of their earthly bodies as a great loss. Diane Piccitto, drawing on Deleuze and Guattari, reads the Vintage decidedly as "an overcoming of 'Eternal Death'" because "operations of becoming where humans are becoming grapes and vice versa" cause "the potency of this event" to lie

> in being in the throes of it and not in some perceived result. Becoming, then, has an element of violence and terror, which is not surprising given that one must open oneself to what is foreign to one's own self, but doing so potentially leads to the path of redemption or fundamental change. (126)

According to the Seven Angels of the Presence, "States Change: but Individual Identities never change nor cease: / You cannot go to Eternal Death in that which can never Die"; "Hell & the Grave!" are "States that are not, but ah! Seem to be" (32:23–24, 29). The "ah!" expresses compassion for how very real they "Seem to be." In the final lines of the description, the Vintage definitely becomes violent, oppressive, and orgiastic. The "Human grapes" suffer "In chains of iron & in dungeons circled with ceaseless fires. / In pits & dens & shades of death: in shapes of torment & woe" tortured by a whole list of instruments: "plates & screws & wracks & saws & cords & fires & cisterns" (27:32–34). The wine-press can use violence to a productive and beneficial purpose; the same could be said of all of these tools. But here the benefit is to the torturers:

> The cruel joys of Luvahs Daughters lacerating with knives
> And whips their Victims & the deadly Sport of Luvahs Sons.

> They dance around the dying, & they drink the howl & groan
> They catch the shrieks in cups of gold, they hand them to one another:
> These are the sports of love, & these the sweet delights of amorous play. (27:35–39)

The enjoyment here does not seem to be sadomasochistic pleasure in pain for both inflictor and sufferer, but sadistic torture, decidedly one-way. The intoxicating blood-wine *is* the pain of the human grapes; it is a harvest of suffering, as the sons and daughters "drink the howl & groan" and "catch the shrieks in cups of gold." The alternate version of this passage in *Vala; or, The Four Zoas* makes the Nettle, Thistle, and Weeds the perpetrators.[4] The tables are turned, with humans being tortured as pruned, nailed, and fired grapes are; the human grapes learn what it is to be "vegetable," their pain and death being for others' pleasure.

Human sexual pleasure seems only a deceptive prelude to the ultimate pain of the Vintage: "Tears of the grape, the death sweat of the cluster the last sigh / Of the mild youth who listens to the lureing songs of Luvah" (27:40–41). This is the reverse of the grapes of sexual arousal grown by Orc's fires in *America*. There they are symbols of reviving desire in the dead, while here it is not feeling the growth of one's own grapes, but the crushing of others' fruit that gives pleasure. There, fire shows its beneficial aspect of warmth, while here it is part of the torture, and, agriculturally, used for the destruction of waste vegetable matter once its yield has been extracted. The purpose of sex seems to be to produce vegetative matter for suffering and death: later in *Milton*, Blake writes the equating phrase, "Mortal & Vegetable in Sexuality" (35:24).

Unlike the flower chorus, the humans in the Vintage are undifferentiated. There is no sense of individual expression; the catalogues are dedicated to the animal, insect, and plant labourers, and their tools of torture. If the Vintage erases the particularity of the humans in the wine-press, then it does not represent cultivation that develops identity through necessary painful limitation; it crosses the "bounding line." However, "Nature has no Outline: but Imagination has" (*The Ghost of Abel* 1). The Vintage may be clearing away undifferentiation itself by crushing the vegetative bodies that obscure humans in Ulro. In this light, the Vintage enacts in an agricultural mode the sacrifice of Milton in his climactic speech: "All that can be annihilated must be annihilated . . . This is a false Body: an Incrustation over my Immortal / Spirit . . . These are the Sexual Garments" to be "purge[d] away with Fire / Till Generation is swallowd up in Regeneration" (40:30, 35–36; 41:25, 27–28). Milton's rhetoric here is such that this speech is generally taken as the poem's exemplary message. However, identity plays a curious part here: this is Milton speaking from Milton's perspective. The very Protestant terms (including "To cleanse the Face of my Spirit by Self-examination" [40:37]) suggest that Milton may not have overcome all of his puritanical errors as he, and Blake, apparently intended. If he had, would he still be Milton?

Milton could be wrong, or at least biased; Los, the director of the Vintage, could be in error too. Los "stood & cried to the Labourers of the Vintage in voice of awe," giving a rousing speech on how the Vintage would proceed and all that it would accomplish, concluding,

> Wait till the Judgement is past, till the Creation is consumed
> And then rush forward with me into the glorious spiritual
> Vegetation; the Supper of the Lamb & his Bride; and the
> Awakening of Albion our friend and ancient companion. (25:59–62)

In response, "lightnings of discontent broke on all sides round" (25:63). This conflict may just be part of the intellectual wars of Eternity; Blake's mythological characters spend much of *Milton* strenuously disagreeing with each other. It may be that Los is over-optimistic about the outcome of the Vintage, and too eager to "rush forward"; his odd idea of "the glorious spiritual / Vegetation" seems suspect if vegetation means benighted material existence. Yet it could be that Los has particular insight into the continuity of all forms of life, vegetable and human, and between material and spiritual worlds. Or is it that, as so often in Blake's prophecies, Los is struggling as best he can with the bad situation that is the fallen world? Considering the price to be paid by the suffering human grapes, and by all creation if it is to be consumed for a contested hope of redemption, it is a terrible thought that Los's plan could be wrong. But what plan could be irrefutably right or free of harm?[5] As Timothy Morton emphasizes, "Any system of interrelated beings . . . is contingent and fragile . . . This means that no ecological action can be complete or absolutely correct" (24).

Conclusion: "the Human Wine stood wondering"

Even though the Vintage is narrated near the end of Book the First, *Milton* ends on a note of suspense with preparations "To go forth to the Great Harvest & Vintage of the Nations" (43:1). Blake refuses to present the Vintage as a completed action; he suspends the outcome in order to confront the violence squarely, insist on the fundamental difficulty of resolving it, and avoid diminishing suffering in the rosy retrospective light of a guaranteed reward. In *The Four Zoas*, Blake boldly depicts the outcome of the apocalypse. It is clearer there that the Vintage is an intended and necessary culmination which partially goes awry. The "sons & daughters / of Luvah quite exhausted with the Labour & quite filld / With new wine . . . began to torment one another and to tread / The weak" (137:18–21). In the following lines, "Luvah was put for dung on the ground by the Sons of Tharmas & Urthona" (137:24). Such an extremely earthy image is surprising to apply to an Eternal; apparently, if humans can be treated as vegetables are in the cycle of growth and harvest, superhuman beings can be treated as manure. A few lines later there is the suggestion, not of punishment, but of the need for Luvah along with his emanation and offspring to go through a cycle of seasons, with the death that brings, before they can be accepted again.

> Luvah & Vala woke & all the sons & daughters of Luvah
> Awoke they wept to one another & they reascended
> To the Eternal Man in woe he cast them wailing into
> The world of shadows thro the air till winter is over & gone. (137:28–31)

The echo of the Song of Solomon ("For, lo, the winter is past, the rain is over and gone; / The flowers appear on the earth" [2:11–12]) is appropriate for the erotic Luvah, and also for the frequent appearance of vines and vineyards in the Song of Solomon among its imagery that likens humans to plants and gardens. As for the human grapes in *The Four Zoas*, the sons of Tharmas and Urthona "formed heavens of sweetest wo[o]d[s] of gold & silver & ivory / Of glass & precious stones," perhaps the best things of the material world, and "loaded all the waggons of heaven / And took away the wine of ages with solemn songs & joy," and "the Human Wine stood wondering in all their delightful Expanses" (137:25–26, 32). Somehow, humans in the form of wine still have the capacity to wonder. During the pressing, in *The Four Zoas* too "the Human Grapes sing not" (136:21), but an essence of them produced by the wine-presses is given a voice, or, rather, a synaesthetic form of expression strikingly similar to that of the flower chorus:

> The blood of life flowd plentiful Odors of life arose
> All round the heavenly arches & the Odors rose singing this song
>
> O terrible wine presses of Luvah O caverns of the Grave
> How lovely the delights of those risen again from death
> O trembling joy excess of joy is like Excess of grief. (135:38–136:3)

They are then described like the Spectres before they are given form: "Forsaken of their Elements they vanish & are no more / No more but a desire of Being a distracted ravening desire" (136:6–7). Unable to take shape of their own efforts—"They plunge into the Elements the Elements cast them forth / Or else consume their shadowy semblance"—they "Cry O let us Exist . . . let us consume in fires / In waters stifling or in air corroding or in earth shut up / The Pangs of Eternal birth are better than the Pangs of Eternal Death" (136:9–15). Unlike Abel's blood that cries unto God of murder (Genesis 4:10), here the voice of blood insists that the suffering is worth its result; or, at least, the released "Odors" can conceive of suffering as redemptive, while the more materially substantial human grapes actually suffering in the press, producing the odours by the breaking of their flesh, only "howl." Peter Otto argues that "the wine press reduces life to elementary fluids (wine / blood), which must then, once again, be drawn into human form" (*William Blake* 748): this is not a release from the material world or its cycles. But neither is it Eternal Death. Agriculture is a feature of the renewed world: the sun "Each morning like a New born Man issues with songs & Joy / Calling the Plowman to his Labour & the Shepherd to his rest" (138:29–30). Either this is a delusive "utopic vision" that is really Urizenic (Otto, *Blake's Critique* 337–338), or the labours and cycles of cultivation remain essential even after the apocalypse.[6]

In *Milton* what follows the Vintage in Book the First is the cultivation of the arts in Golgonooza, and the cultivation of Spectres like seeds or artworks: again there is the possibility that the apocalypse is circular, bringing us back to vegetative life, but the added emphasis on cultivation in the aesthetic sense involves art in the suspended question of redemption. Within the description of the Vintage in *Milton* are these lines, absent from the *Four Zoas* version:

> This Wine-press is call'd War on Earth, it is the Printing-Press
> Of Los; and here he lays his words in order above the mortal brain
> As cogs are formd in a wheel to turn the cogs of the adverse wheel.
> (27:8–10)

There is violence in the printing press too, which was originally modelled on the wine-press. Like the wine-press it applies force to vegetable matter (paper, in Blake's day made of recycled linen clothing made of flax) to transform it. This transformation can equally be seen as the cultivation or levelling of identity, since the printing press is a tool to give form and being to an author's or artist's ideas, yet, as Otto writes, it "performs the same set of operations no matter what is placed within it. For the printing press it is of no consequence whether it prints the works of Burke or Paine" (*Constructive Vision* 83). As "Los puts all into the Press, the Opressor & the Opressed" (25:6), Blake seems to consider indiscriminacy inherent in the word, also used for press-gangs that arbitrarily force men into military service. The printing press is historically implicated as a catalyst for war, and what is "call'd War on Earth" is a prime example of justification of violence and suffering for their supposed benefit or ideological necessity. *Milton* contrasts intellectual war in Eternity with corporeal war on Earth,[7] but both are war; there is no way to exonerate either from error and violence, nor deny that either can be a terrible but powerful agent of change. Is the printing press, like the wine-press, a torture instrument? Reading can be painful, causing physical discomfort (eyestrain, headaches) and requiring mental discipline, not only in first acquiring the skill, but perpetually in the case of demanding works, and Blake's works are notoriously demanding. Mary Lynn Johnson describes how *Milton* causes readers to "find themselves clueless *in medias res*, thrown back on their own resources, as if subjected to a wilderness-survival test" (231). Los's press, with its wheels and cogs, is the same kind of rolling press used by Blake (Essick and Viscomi 175). If Los is potentially in error using violent means, then so is Blake. But it may be that error and correctness, when it comes to Los's and Blake's apocalyptic efforts, can be as resistant to differentiation as the wild and the cultivated, and as equally, inextricably necessary to the work of creation and the possibility of redemption.

Notes

1 All references to Blake's writing are taken from Erdman's edition. Verse quotations are cited by plate or page and line number.

2 Eaves, Essick, and Viscomi date the composition of *Milton* to 1804–c.1811, with additional plates made probably up to 1818.

3 There are rare exceptions in variations on the formula animal, vegetable, mineral (*Milton* 25:21; *Descriptive Catalogue* 567; *Chaucer* 589).

4 Eaves, Essick, and Viscomi date *The Four Zoas* to c.1796–1807, largely earlier than *Milton* but partly overlapping.

5 Hobson writes, "Los's labours *are* redemptive, but in a mixed, contradictory way . . . labour and imagination have created *all* human culture, necessarily an oppressive culture, and yet are the only forces that can struggle to transform that culture" (260–261).

6 For a sense of how "diametrically opposed" readings of the apocalypse in *The Four Zoas* can be, see Hobson (203–204).

7 The most strident statements on the subject are in the Preface, included in only two copies of *Milton*.

Works cited

Bentley, G. E., Jr. *Blake Records*, 2nd ed. Yale UP, 2004.

Blake, William. *The Complete Poetry and Prose of William Blake*, edited by David V. Erdman. U of California P, 2008.

Broglio, Ron. "Incidents in the Animal Revolution." *Beyond Human: From Animality to Transhumanism*, edited by Charlie Blake, Claire Molloy, and Steven Shakespeare. Continuum, 2012, pp. 13–30.

Bruder, Helen P. and Tristanne Connolly. "Introduction: 'Conversing with the Animal forms of Wisdom.'" *Beastly Blake*, edited by Helen P. Bruder and Tristanne Connolly. Palgrave Studies in Animals and Literature. Palgrave Macmillan, 2018, pp. 1–35.

Eaves, Morris, Robert N. Essick, and Joseph Viscomi. *The William Blake Archive.* 2019, www.blakearchive.org

Eliot, T. S. "Blake." *The Sacred Wood: Essays on Poetry and Criticism.* 1920. Methuen, 1960. 151–158.

Erdman, David V. *A Concordance to the Writings of William Blake*, 2 vols. Cornell UP, 1967.

Essick, Robert N. and Joseph Viscomi, eds. *Milton a Poem and the Final Illuminated Works.* Blake's Illuminated Books Vol. 5. Tate, 1993.

Hobson, Christopher Z. *The Chained Boy: Orc and Blake's Idea of Revolution.* Associated UP, 1999.

Hutchings, Kevin D. *Imagining Nature: Blake's Environmental Poetics.* McGill-Queen's UP, 2002.

Johnson, Mary Lynn. "*Milton* and Its Contexts." *The Cambridge Companion to William Blake*, edited by Morris Eaves. Cambridge UP, 2003, pp. 231–250.

Kennedy, John. *A Treatise upon Planting, Gardening, and the Management of the Hot-House*. 2nd edn. 2 vols. S. Hooper, 1777.

Miller, Philip. *The Gardeners Dictionary*, 8th ed. 3 vols. Printed for the author, 1768. N. pag.

Milton, John. *Paradise Lost. The Complete Poetry of John Milton*, edited by John T. Shawcross. Doubleday, 1971, pp. 249–517.

Morton, Timothy. "Queer Green Sex Toys." *English Language Notes*, vol. 54, no. 1, 2016, pp. 15–26.

Otto, Peter. *Blake's Critique of Transcendence: Love, Jealousy, and the Sublime in The Four Zoas*. Oxford UP, 2000.

———. *Constructive Vision and Visionary Deconstruction: Los, Eternity, and the Productions of Time in the Later Poetry of William Blake*. Clarendon P, 1991.

———, ed. *William Blake*. 21st-Century Oxford Authors. Oxford UP, 2018.

Piccitto, Diane. "Apocalyptic Visions, Heroism, and Intersections of the Human and 'the Not Human' in Blake's *Milton*." *Beastly Blake*, edited by Helen P. Bruder and Tristanne Connolly. Palgrave Studies in Animals and Literature. Palgrave Macmillan, 2018, pp. 111–133.

6 Wild plants and wild passions in Percy Bysshe Shelley's poems for Jane Williams

Cian Duffy

Mary Shelley did not like pumpkins. In a letter of 2 November 1818, she complains to her friend Maria Gisborne about "the streets" of Venice, where she had just been, "because they are narrow and dirty, and above all because they carry zucche about to sell, the sight of which always makes me sick" (*Letters* 1: 81). It is therefore surprising that Mary's husband, Percy Bysshe Shelley, later chose to make this particular plant the focus of an unfinished poem, "The Zucca," in which he probably intended to address his sexual attraction to their mutual friend, Jane Williams.[1] However, "The Zucca," begun in the winter of 1821–22, is only one of a number of poems occasioned by Shelley's feelings for Jane in which he uses wild plants as figures for exploring wild passions. Other examples include "Remembrance," "The Magnetic lady to her patient," and the so-called "Unfinished Drama" which Shelley began to compose for the members of his circle at Pisa in the spring of 1822. These poems represent the culmination of what we might call Percy Shelley's *botanical poetry*: poems which not only have a plant as their ostensible subject, or which develop extended plant imagery, but which also engage, either explicitly or implicitly, with contemporary botanical discourses and practices. Other examples of this genre, such as Shelley's "Verses written on receiving a Celandine in a letter from England" (1816) and "The Sensitive-Plant" (1820), have been relatively well examined by editors and critics of his work. But Shelley's botanical verse dealing with Jane Williams—and "The Zucca" and the "Unfinished Drama" in particular—is still comparatively neglected. Hence, defining and appraising Shelley's late botanical poetry addresses an under-studied aspect of his writing about the natural world and potentially sheds light on his relationship with Jane Williams. Doing so within the covers of a volume dealing with "wild Romanticism" is especially appropriate because of the relationships which Shelley's late botanical poetry forges between different kinds and modes of wildness, between the wildness of nature and the wildness of human emotions. These poems take up and transform the tropes of contemporary botanical discourse and, in so doing, further evidence the extent to which Shelley's nature poetry responds to contemporary debates in natural philosophy. The anthropomorphic analogies at the core of Shelley's late botanical poems, between wild plants and wild passions, also set them apart from much of Shelley's more familiar writing about nature, which tends to emphasize rather the ontological distinctness of the natural world.

"Books & plants about me": Shelley and botany

In a letter of 11 January 1822, written around the time he was working on "The Zucca," Shelley told his friend Thomas Love Peacock that he had "collected books & plants about me" in his study at Pisa and that "our windows are filled with plants which turn the sunny winter into spring" (*Letters* 2: 373–374). Such domesticated greenery contrasts notably with the wild places and wild phenomena—with the alps, earthquakes, storms, and volcanoes—which so often take centre stage in Shelley's poetry. But Timothy Morton does well to remind us that Shelley "was also interested" in flora and fauna: "in animals, plants, and what humans do to them and with them" ("Nature and Culture" 193). Morton has written extensively about Shelley's vegetarianism and about his contributions to various other late eighteenth- and early nineteenth-century discourses around food, diet, and what we would today call "animal rights." But Shelley's interest in plants and botany has yet to receive sustained attention from scholars of his work. Marilyn Gaull is correct to note that botany was not amongst "the sciences that most interested Shelley": it does not have the same place in his work as "geology, astronomy, chemistry, biology" (588). However, Shelley's notebooks, correspondence, poetry, and prose all reveal a lifelong interest in plants as well as an awareness of and an engagement with botanical discourses and practices.

By the end of the eighteenth century, botany had established a significant cultural presence in Britain.[2] Joseph Banks, president of the Royal Society, had gradually expanded the Royal Botanic Gardens at Kew into one of the leading collections of domestic and exotic specimens in the world, and diverse efforts were underway to produce systematic taxonomies of British flora. Prominent examples here include the *Flora Londinensis* (1777) compiled by William Curtis and James Sowerby, and the monumental, thirty-six-volume illustrated *English Botany* (1790–1813) produced by Sowerby himself, with textual commentaries provided by James Edward Smith, founder of the Linnean Society of London. Mary Wollstonecraft studied botanical drawing with Sowerby. Nor were such taxonomic efforts by any means restricted to the flora of Britain alone. The plants of India, in particular, became an especial focus for British botanists and colonial administrators in the late eighteenth century. William Jones, for example, the founder of the Asiatic Society of Bengal, whose *Collected Works* Shelley ordered in December 1812, authored a "Design for a Treatise on the Plants of India" (1794), and comprehensive, illustrated compendia of Indian flora published during Shelley's lifetime include *Plants of the Coast of Coromandel* (1795–1813), *Horta Bengalensis* (1814), and the posthumous *Flora Indica* (1820) compiled by the Scottish surgeon William Roxburgh.

As Theresa Kelley and a number of other scholars have made clear, the identification, classification, and circulation of plants across national boundaries were deeply implicated in the cultural politics of empire. "By the end of the eighteenth century," Kelley explains, botanical studies function as "strategic mechanisms" to "possess" through codification the "botanical plentitude" of "distant territories"—effectively, to use the terms of today's ecological discourses, *dewilding* the wild ("Romantic Exemplarity" 226). "Bringing exotic plants home to England to make them grow and bloom . . . or transporting English plants to distant colonies," whether for scientific or for commercial reasons, was a means of "domesticating an exotic plenitude that was by turns strange and strangely familiar" (Kelley, "Romantic Exemplarity" 227). One of the more striking examples of this process is the lavishly illustrated *Temple of Flora* (1799–1807) compiled by Robert Thornton—to which, as we shall see, Shelley's late botanical poetry may owe a particular debt—in which domestic and exotic flora are presented side-by-side as part of an implicitly British "garden of the botanist, poet, painter and philosopher" (as Thornton's subtitle has it).

Moreover, as a branch of natural philosophy, the study of plants during Shelley's lifetime tended to cross frequently what we would today think of as the disciplinary boundaries between different areas of enquiry and genres of writing. The hypothesis formulated by Goethe, for example, in his *Versuch die Metamorphose der Pflanzen zu Erklären* (1790), that different leaf shapes imply the existence of a primordial plant (*urpflanze*) from which all others derive, typifies just how quickly the examination of plants could lead to larger-scale theorizing about the nature and origins of organized life. One consequence of this pre-disciplinary character of botany was that botanical enquiry could often be invoked as a figure for other, ostensibly unrelated kinds of enquiry. Hence, "when romantics—poets, botanists, philosophers, and assorted others—thought about plants, they did so," as Kelley puts it, "across several registers": botany, during Shelley's lifetime, involved "a disciplinary and cultural array of practices" which "operated across romantic culture" (*Clandestine Marriage* 22).

Shelley engaged with this "disciplinary and cultural array" in a variety of ways and on a variety of levels. Shelley had, for instance, at least an amateur interest in gardening. We have already noted the plants which Shelley kept in his lodgings in Pisa in 1821–22. Earlier, in July 1816, Shelley wrote to Peacock from Chamonix asking him to find him a house "near" Windsor Forest, "with as good a garden as may be" (*Letters* 1: 489). And in a follow-up letter written a week later, on 25 July, Shelley told Peacock that he had purchased "a large collection of the seeds of rare Alpine plants, with their names written upon the outside of the papers" (*Letters* 1: 501). "These," Shelley says, "I mean to colonize in my garden in England; & to permit you to make what choice you please from them" (*Letters* 1: 501). Whilst travelling in Europe and, later, whilst resident in Italy, Shelley often sent and received plants in correspondence with friends in England. On a number of occasions, this led to botanically themed poems or images, as in the "Verses written on receiving a Celandine in a letter from England," which Shelley composed at Chamonix in 1816, having received the flower in an earlier letter from Peacock (see Blank 56–63; Quinn 88–109). Shelley's well-known lyric beginning "The flower that smiles to-day" was almost certainly prompted by his receipt in October 1822 of a "little flower from Marlow" in a letter sent to Pisa by Hogg (*Letters* 2: 359–60n.1). Shelley was knowledgeable enough to recognize Hogg's "little flower" as "Milkwort" (*polygala vulgaris*), and in his poem, he uses its faded condition as an analogy for the transience of human joy: "The flower that smiles to-day / Tomorrow dies; / All that we wish to stay / Tempts and then flies" (lines 1–4).[3] Shelley also took Hogg's letter as the opportunity to signal (somewhat facetiously) his approval for what we might now call the ethics and the cultural politics of botany. "As to botany," Shelley writes, "how much more profitable & innocent an occupation it is than that absurd & unphilosophical diversion of killing birds" (*Letters* 2: 361). Shelley's notebooks are full of sketches of plants: some mere doodles to keep pen on paper whilst composing; others much more detailed observations of trees and flowers, often interspersed with fragments of verse or notes. And of course Shelley's letters, poetry, and prose are replete with localized floral imagery, such as the beautiful star-flower metaphors which we find in *Adonais* (1821) and "The Triumph of Life" (1822), or the extended description of the flora thriving amidst the ruins of Rome in Shelley's short story, "The Coliseum" (1819).

There is ample evidence, then, for Shelley's interest in plants and in botanical discourses and practices—and it is against the backdrop of this evidence that I want, now, to examine Shelley's use of botanical imagery in "The Zucca," "The Magnetic lady to her patient," and the "Unfinished Drama," and the figurative analogies which he establishes, in those poems, between the wildness of plants and the wildness of human emotions.

"The Exotic as you please to call me": botany in the Jane Williams poems

Shelley began work on "The Zucca" around the turn of 1821–22. Manuscript evidence confirms this, as do the references in the final lines of the draft to the "savage storm / Waked by the darkest of December hours," which presumably reflects the stormy weather recorded at Pisa by Mary and Percy in letters written during the last week of December 1821.[4] Shelley's draft extends to eleven stanzas of ottava rima before breaking off, though not all the stanzas are complete, and the whole draft is heavily worked; it is not clear how much more Shelley intended to write, but the draft, which breaks off as it begins to describe the effects of winter on the natural world, seems far from finished. A text based on this draft was first published by Mary Shelley in *Posthumous Poems* (1824) but "The Zucca" has been omitted from many subsequent editions of Shelley's poetry. Weinberg is right that it "might be regarded as the most elusive of [Shelley's] Italian fragments," and it remains comparatively neglected by scholars (305).[5]

Wildness takes centre stage in "The Zucca," both the wildness of the plant itself and the wildness of the human emotions of which that plant is made the emblem. And it is worth remembering, as I said at the beginning of this essay, that there is something *wild* in the very conception of the poem: in Shelley having chosen as the focus for a poem in which he engages with his feelings for another (married) woman the very plant whose fruit made his wife feel sick. "The Zucca" opens with a distraught speaker suffering in "infant Winter" from a kind of existential crisis, which he couches in the language of frustrated, erotic longing—"desiring / More in this world than any understand"—and left, in consequence, with a "lorn heart" which can only "weep / The instability of all but weeping" (lines 2, 3–4, 6, 10). In the third stanza, this speaker attempts to explain what exactly he is "desiring" with such devastating consequences:

> I loved . . . oh no, I mean not one of ye
>   Or any earthly one, —though ye are dear
> As human heart to human heart may be;
>   I loved, I know not what, but this low sphere
> And all that it contains, contains not thee:
>   Thou, whom seen nowhere I feel everywhere. (lines 17–22)

One wonders to whom ("ye") the speaker addresses this explanation: to a singular reader, using an archaic appellation or to a group of people whom Shelley meant to identify later in the poem, or perhaps implicitly those making up Shelley's circle at Pisa? In any case, this passage certainly also recalls similar laments in Shelley's quasi-autobiographical poems "Hymn to Intellectual Beauty" (1816) and *Epipsychidion* (1821), where a speaker regrets the "inconstant glance" of some unspecified "Spirit of Beauty" which "doth consecrate" and then "leave . . . vacant and desolate" both the individual "human heart" and the "vale of tears" which all humans inhabit ("Hymn," lines 6, 13, 16–17).

At this point in "The Zucca," in the sixth stanza, the plant is introduced by the disconsolate speaker, who discovers it whilst he is walking alone by a river:

> And thus I went lamenting, when I saw
>   A plant upon the river's margin lie,
> Like one who loved beyond his nature's law
>   And in despair had cast him down to die.
> Its leaves which had outlived the frost, the thaw
>   Had blighted, as a heart which [ ] eye
> Can blast not, but which Pity kills. The dew
> Lay on its spotted leaves like tears too true. (lines 41–48)

As Weinberg observes, after the title, the zucca is only ever referred to "neutrally or impersonally" in the extant draft "as 'the plant' or 'it,'" which, Weinberg says, "endows the mundane plant with an emblematic quality that . . . incorporates, and extends beyond, its biological self" (305). However, the startling anthropomorphism used by the speaker when he first sees the plant, which is extended later in the poem when the speaker attributes to the plant something like a cardiovascular circulatory system, is evidently a reflection of his own state of mind. For he, too, has "loved beyond his nature's law" by "desiring / More in this world than any understand" (lines 3–4). The ailing wild plant images the speaker ailing from the consequences of his wild desires.

In the remaining stanzas (eight to eleven) of the draft, the speaker describes how he carries the plant back to his "chamber" and cares for it:

> The mitigated influences of air
>   And light revived the plant, and from it grew
> Strong leaves and tendrils, and its flowers fair,
>   Full as a cup with the vine's burning dew,
> O'erflowed with golden colours, an atmosphere
>   Of vital [   ] enfolded it anew
> And every impulse sent to every part
> The unbeheld pulsations of its heart. (lines 59–66)

The radical chemist Joseph Priestley, whose work Shelley knew well, had begun in the 1790s to study the process which Charles Reid Barnes would later name photosynthesis. Shelley could also have read in the entry on "Botany" in his copy of the *British Encyclopedia* (1809), which he ordered in December 1812 (see *Letters* 1: 343), how "air and moisture and light have considerable and even the most important effect upon the leaves of plants" and that "sap is carried into the leaves for the purposes of being acted on by air and light, with the assistance of moisture." In "The Zucca," however, the revival of the plant is due to more than renewed photosynthesis in a "mitigated" environment. In fact, in the last two stanzas of the draft, the recovery of the plant (and its anthropomorphic "heart") is paralleled and partly enabled by the recovery of the speaker (and his "lorn heart"), further cementing what Weinberg calls "an allegiance . . . which points to self-identification" on the part of the speaker (306). More precisely, the zucca revives not just under the speaker's care, but specifically because the speaker sheds tears upon it:

> Well might the plant grow beautiful and strong
>   Even if the air and sun had smiled not on it,
> For one wept o'er it all the winter long
>   Tears pure as Heaven's rain, which fell upon it
> Hour after hour. (lines 67–71)

The speaker is moved to these tears by the "gentle lips" of an unidentified woman singing to her own accompaniment "sounds of softest song / Mixed with stringed melodies" (lines 71–73). The speaker's suffering is relieved by these tears, which "loosed his heart," just as the ailing plant is revived by the tears the speaker sheds—whilst the woman's singing, ultimately, redeems them both (line 75). At this point, Shelley's draft breaks off, just as it begins to contrast the domestic setting with the winter landscape outside.

In "The Zucca," then, Shelley sets up a complex sequence of analogies in which the suffering triggered in the speaker by his wild but unfulfilled desires is imaged as a dying wild plant. The "blighted" plant is the type of the "lamenting" speaker, and the means of his recovery—the tears which he sheds in response to the singing of the woman—is also the means by which the plant recovers. Their fates, and their salvations, are, it seems, analogous. These analogies become considerably more complicated, however, once we begin to map the various autobiographical resonances which Shelley seems to have encoded in "The Zucca." To begin with, the "lorn heart" of the speaker matches very well Shelley's own deep unhappiness during the winter of 1821–22: the consequence of his sense of having failed as an author, especially in comparison to his friend Byron; the death of all but one of his children with Mary, and the crisis in their marriage; his despair over the course of British and European politics—and, of course, his frustrated desire for Jane Williams. In a letter to Claire Clairmont of 11 December 1821, Shelley sums up this malaise whilst apparently accepting Claire's earlier designation of him as a species of plant: "The Exotic as you please to call me," Shelley writes, "droops in this frost—a frost both moral & physical—a solitude of the heart" (*Letters* 2: 367). The terms of this letter certainly invite comparison with the "blighted" plant of the "The Zucca," suggesting that Shelley not only intended that plant to function as a type of the speaker in the poem but also for both plant and speaker to function as types of *himself*. And this suggestion is strengthened by another of Shelley's poems occasioned by his relationship with Jane Williams, "The Magnetic lady to her patient," in which Shelley again casts himself as an ailing plant.

In this poem, which is broadly contemporary with "The Zucca," a woman, based on Jane Williams, hypnotizes a man, based on Shelley, in the hope of alleviating various physical and emotional sufferings, with the latter springing explicitly from his unrequited desire for her.[6] Having assured her patient "I love thee not" and urged him to "forget me, for I can never / Be thine," the woman proceeds to offer "pity on thy heart": "Like a cloud big with a May shower," she says, "My soul weeps healing rain / On thee, thou withered flower" (lines 4, 10, 26–27, 28–30). In "The Magnetic lady," then, we find not only another instance of Shelley representing himself as a suffering plant, but also a complex figure in which a woman heals a suffering person by metaphorically weeping upon him. An earlier version of this image is found in Shelley's lyric "On a Dead Violet," which he composed in 1820 after a brief infatuation with Sophia Stacey. *Viola odorata* is traditionally associated with love and desire, as in, for example, *The Loves of the Plants* (1791) by Erasmus Darwin, where it is described as "lovesick" (1: 13). In Shelley's poem, an "abandoned" lover weeps over "the shrivelled, lifeless, vacant form" of the dead flower, the image of departed passion (lines 5–6). And a more troubling precursor for the image can be found in the gothic-botanical narrative poem "Isabella, or The Pot of Basil" (1818), by John Keats, which Shelley read shortly after it was published: Isabella weeps over a pot of basil in which she has placed the severed head of her former lover, Lorenzo (see *Letters* 2: 239, 244). Equally, just as the lady who "weeps healing rain" on her suffering, would-be lover in "The Magnetic lady" is based on Jane Williams, so is it tempting to read the "gentle" woman whose singing triggers the speaker's healing tears in "The Zucca" as also a representation of Jane, whose musical abilities Shelley often praised.

What I have been suggesting, then, is that Shelley's use of botanical imagery in "The Zucca" sets up complex relationships between different kinds of wildness. The wild plant serves as a figure for the speaker, the effect of whose wild desires is mirrored in the "blighted" condition of the plant. But both the speaker and his self-image in the plant also serve, apparently, as figures for Shelley himself, a self-proclaimed "Exotic" who is also suffering a "moral & physical" "frost" in consequence of having "loved beyond his nature's law" in desiring Jane Williams. Such a complex poetic situation also raises (at least) two questions. First: could Shelley find any solutions to these problems of unrequited desire, either as formulated in "The Zucca" or in real life? The fact that Shelley's draft breaks off in mid-winter suggests that he could not, but the answer is not quite so simple. The second question is actually the easier of the two. Why might Shelley have chosen a botanically themed poem to address these actual and imagined problems of desire, beyond merely having taken a helpful hint from Claire Clairmont? Why use a wild plant as a figure for exploring wild passions? To answer this question, we must return to Shelley's engagement with contemporary botanical discourse.

"A mournful type of thee": botany and analogy in the Romantic period

One of the earliest instances in which Shelley uses a wild plant to figure another, ostensibly unrelated form of wildness occurs in "To Harriet *****" (1812), in which Shelley dedicates his first major poem, *Queen Mab* (1813), to his first wife, Harriet Westbrook. "Thine are these early wilding flowers," Shelley writes, "though garlanded by me" (lines 11–12). On a first reading, these lines might seem like nothing more than a conventional apology for the roughness of the verse (its "wildness") now being offered as a token of devotion ("garland") to the beloved. However, Shelley's image is more complex, as becomes apparent if we examine what he actually meant by "wilding." In late eighteenth- and early nineteenth-century botanical discourse, as well as in general use, *wilding* could refer either simply to uncultivated ("wild") plants or, more specifically, to wild plants descended from cultivated ones. Hence Shelley's figure is less a metaphor than an *analogy*; less an apology for the quality of the verse than a valorization of the formal innovation of the poem, which blends styles and genres, as well as of its "wild" iconoclastic content.[7] As Sam George (15, 153–174) has shown, British botanical writing during the Romantic period tended to praise indigenous wild flowers as more robust and more beautiful than cultivated, exotic flowers—and that is the point that Shelley is making with his *wilding* analogy. *Queen Mab* is uncultivated in that it departs from the restrictive norms of content and form to pursue a revolutionary agenda: the *wildness* of the poem, in short, is what we might now call *Romantic*.

Shelley would have found many precedents for this kind of analogy between wild flowers and other kinds of wildness in contemporary botanical writing. As Kelley, Maniquis, and others have noted, *analogy* actually played a key role in the more speculative elements of eighteenth-century botanical enquiry. Of particular interest was the question of where the boundary between plant and animal life might be drawn and of how, exactly, the difference between plant and animal life might be defined. As Maniquis observes, plant-animal analogies "gained more and more elaboration" in eighteenth-century natural philosophy and by the early nineteenth century, botanists "could see the likeness in almost every way, plants now assuming the processes of respiration, digestion, generation, and their own special form of muscle structure" (135)—and we remember in this regard the apparently-incongruous cardiovascular system of the plant in Shelley's "The Zucca."

Shelley was well aware of such debates. He wrote to Hogg in October 1821 describing some "curious fleshy flowers" found around Pisa—presumably succulents—"& one that has blood & that the peasants say is alive" (*Letters* 2: 361). The most-familiar instance of a botanical analogy in Shelley's verse is to be found, of course, in "The Sensitive-Plant" (1820), which has by now been very well studied.[8] As Maniquis notes, *mimosa pudica*—the so-called "sensitive plant"—became a major focus of botanical enquiry during the eighteenth century because of its rapid reactions to changes in the level of light and to direct physical stimuli, raising questions about whether it was a plant, an animal, or some hybrid of the two. *Mimosa pudica* is described in detail by Darwin in the *Loves of the Plants* and has its own entry in Nicolson's *British Encyclopedia* (see Figure 6.1).

Shelley's poem describes a *mimosa pudica* which grows along with many other cultivated and wild plants in a garden tended by "a Lady: the wonder of her kind, / Whose form was upborne by a lovely mind" ("Part Second," lines 5–6). As George notes, the "trope" of a woman tending a garden was not uncommon in eighteenth-century botanical writing and became more frequent as women became more involved with the practice of botany and as women's writing increasingly addressed botanical topics; it is "central,"

Figure 6.1 "Mimosa Grandiflora" from Robert John Thornton, *Temple of Flora* (1812); reproduced by kind permission of The Cleveland Art Museum (Wikimedia Commons).

as George points out, to one of Shelley's favourite novels: *Julie, or The New Heloise* (1761), by Jean-Jacques Rousseau (6, 8–11). Shelley's "Lady" certainly invites comparisons with the figures based on Jane Williams in "The Zucca," "The Magnetic lady," and "Unfinished Drama," and indeed, in a letter to Leigh Hunt of 19 June 1822, Shelley calls Jane "a most delightful person . . . whom we all agree is the exact antitype of the lady I described in the Sensitive plant—though this must have been a *pure anticipated cognition* as it was written a year before I knew her" (*Letters* 2: 438). Shelley's *mimosa pudica* thrives under the Lady's care, but once she dies, "the garden once fair became cold and foul / Like the corpse of her who had been its soul" ("Part Third," lines 17–18). Weeds survive this winter and when spring returns, "the mandrakes and toadstools and docks and darnels, / Rose like the dead from their ruined charnels" ("Part Third," lines 116–117). But the

"Sensitive-Plant," the type of the Lady by whose care it was nurtured, is reduced to a dead "wreck" ("Part Third," line 115).

As Desmond King Hele (209–210) first noted, the treatment of *mimosa pudica* in Darwin's *Loves of the Plants*, which ends with an emotive description of the death of the plant in winter, is an important stylistic and conceptual hypotext for Shelley's poem. Darwin's account, which Shelley knew well and which is also quoted in the descriptions of *mimosa* in Nicholson's *British Encyclopedia* and Thornton's *Temple of Flora*, relies heavily on personification. So does Shelley's poem, but as Kelley observes, Shelley "expands the reach of Darwinian plant personification by teasing out figurative possibilities that move from a paradise of animated plants to its negation, a garden animated by cranky, noxious weeds" (*Clandestine Marriage* 225). One such "figurative possibility" is to read the transition from cultivated garden to wilderness following the death of the Lady as an allegory for the collapse of civil society in the absence of beneficent government. But Shelley also moves in his "Conclusion" to "The Sensitive-Plant" from personification to analogy, by configuring the *mimosa* and the Lady as types of each other. And, as Maniquis first argued, the point of this analogy is to enable speculation not about plants but about the "limited" nature of *human* perception (145). In Shelley's "Conclusion," the speaker of "The Sensitive-Plant" wonders whether either the dead plant or the dead lady has any continued existence. The speaker concludes that whilst it is impossible to answer these questions, it is a "pleasant" and a "modest creed" to "own that death must be, / Like all the rest—a mockery" and that "for love, and beauty, and delight / There is no death nor change" ("Conclusion," lines 13–16, 21–22). In this final part of "The Sensitive-Plant," then, the analogy between *mimosa* and Lady becomes, as Maniquis argues, an analogy between *mimosa* and human beings. Just as the *mimosa*, for all its sensitivity, has no real understanding of the nature of the physical stimuli which determine the conditions of its existence, so do we humans have no real understanding of the ontological conditions of *our* existence. In "this life" which is dominated by "error, ignorance, and strife," Shelley's poem concludes, the true nature of reality "exceeds our organs—which endure / No light, being themselves obscure" ("Conclusion," lines 23–24).

Botanical personification and analogy are central, then, to "The Sensitive-Plant," although in the final analysis, Shelley uses the highly detailed and informed botanical material to enable consideration of non-botanical matters, including poetics, politics, and ontology. The wild is *dewilded* for rhetorical purposes; the botanical serves to make a point about the human. We have already discerned a similar strategy in "The Zucca," where the ailing wild plant makes possible a rhetorical engagement with the wild passions of the speaker and, apparently, of the author. In "The Zucca," however, Shelley's plant-person analogy is not just physiological or even psychological: it is actually anthropomorphic, as well as potentially autobiographical. But in developing such a specialized plant-person analogy in order to examine the relationship between existential and erotic desire, Shelley was also, I want to suggest, taking up and transforming one of the key tropes of contemporary botanical writing: the *sexualized* plant.

During Shelley's lifetime, the theory and praxis of botany in Britain was still dominated by the work of Carl Linnaeus, whose writings were widely available in English—in translation, in the responses of British botanists, and in such literary remediations as Darwin's *Loves of the Plants* and William Mason's *The English Garden* (1786). In his *Philosophia Botanica* (1751) and *Species Plantarum* (1753), Linnaeus applied to the classification of all plants then known the system of binomial nomenclature which he had been developing since the publication of *Systema Naturae* in 1735. "Linnaeus," Shelley would have read in the entry on "Botany" in the *British Encyclopedia*, was "the first person who took a very comprehensive and philosophical view of the laws of system, and at the same time carried them most happily into effect."

The Linnean method of classifying natural phenomena undoubtedly constituted what we would now call, following Thomas Kuhn, a paradigm shift in the natural sciences during the eighteenth century. It played a key role in the transition from the pre-disciplinary practice of natural philosophy, inherited from the early modern period, towards the evidence-based practice of natural history which became the foundation for the emergence, in the early nineteenth century, of distinct, semi-professional scientific disciplines (see Heringman 3, 4, 6–7). However, despite Linnean botanical classification serving as a methodological "bellwether," as Heringman calls it, for the transition from natural philosophy to natural history, botany, as a set of institutional and cultural practices, paradoxically retained "its proto-disciplinary complexity longer than the other branches of natural history" (11).

This persistent "complexity"—the tendency of botany to become bound up with other, ostensibly unrelated areas of enquiry and genres of cultural productivity—was due to a very considerable extent to the decision taken by Linnaeus to make the sexual characteristics of plants the basis for his nomenclature. The Linnean system, as George explains, specifically "focused attention on the organs of generation and was termed the 'Sexual System,'" whilst Linnaeus, in his writings, "famously made use of human-plant analogies; his nomenclature was inspired by traditional wedding imagery and marriage metaphors permeate his botanical taxonomy" (1). The sexual focus of Linnean botany, and the eroticized language in which it was routinely encoded, reached its apotheosis in Darwin's *Loves of the Plants* which "makes explicit," as George observes, "the language used to describe the marriage of plants in Linnean texts in the eighteenth century" (2). Shelley, as we have seen, knew Darwin's poem well, and in his letter to Hogg of 22 October 1821, Shelley makes use of the Linnean term "the Cryptogamia genus" to describe some of the local flora "which I had never remarked in England—ferns especially" (*Letters* 2: 361). Should any further evidence for Shelley's awareness of the sexual charge of Linnean botany be required, we might remember the incestuous twist which he gives the Linnean trope of botanical marriages in "Love's Philosophy" (1820), another playfully erotic poem occasioned by Shelley's infatuation with Sophia Stacey, which insists that "No sister-flower would be forgiven / If it disdained its brother" (lines 11–12).

Hence the issue which often confronted those practising Linnean botany during the Romantic period was not only that botanical discourse routinely involved "slipping from plants to humans," as Kelley puts it, but moreover that "thinking about plants invites thinking about sex" (*Clandestine Marriage* 117)—and this made Linnean botany controversial throughout Shelley's lifetime. In his poem *The Unsex'd Females* (1798), for example, Richard Polwhele deemed Linnean botany to be an inappropriate pursuit for women precisely because of the close literal and figurative emphasis on sexuality in botanical texts and practices. Even the rather more progressive William Jones, who insisted in his "Design for a Treatise" that the Linnean "distribution" would "ever be found the clearest and most convenient of methods," baulked at "the allegory of *sexes* and *nuptials*": this, Jones says, "ought to be discarded, as unbecoming the gravity of men, who, while they search for truth, have no business to inflame their imaginations" (3: 4, 5–6).

As Quinn (88–109) and Blank (56–63) first observed, Shelley uses a dead plant to figure the moral death of an identifiable individual in his "Verses written on receiving a Celandine in a letter from England," which he composed in Chamonix in July 1816, having received a desiccated *ficaria verna* in the post from Peacock. In this case, the target was William Wordsworth, whose youthful radicalism Shelley admired, but whom he had recently criticized, in his sonnet "To Wordsworth," for having betrayed the commitment to "truth and liberty" (line 12) which Shelley found in *Lyrical Ballads* (1798). In Shelley's richly intertextual "Verses written," the "withered" and "deformed" celandine which reached him in Chamonix is configured as a "mournful type" both of its own living self and of Wordsworth, who had written about *ficaria verna*, and adopted it as his own flower in two lyrics first published in *Poems in Two Volumes* (1807): "To the Small Celandine" and "The Small Celandine."

However, Shelley's "Verses" also equate Wordsworth's moral collapse with a failure of passion, thereby introducing a specifically Linnean element to the botanical personification which has not yet been noted by critics: "he is changed and withered now, / . . . his heart is gone" (lines 29–31), Shelley writes, anticipating the "blighted" and "lorn" hearts of the plant and the speaker in "The Zucca." In describing Wordsworth in such terms, Shelley anticipates the explicitly sexual terminology of his later critique of Wordsworth's personal and artistic failures in "Peter Bell the Third" (1819), where Wordsworth is portrayed as "a kind of moral eunuch; / He touched the hem of Nature's shift, / Felt faint—and never dared uplift / The closest, all-concealing tunic" ("Part Fourth: Sin," lines 314–317).

But if Wordsworth's perceived apostasy stems, in Shelley's analysis, from a failure of passion, then the speaker of "The Zucca" is, conversely, the victim of too wild a passion, of having "loved beyond his nature's law." In Linnean botany and its many literary appropriations, Shelley would often have found sexualized and eroticized personifications of plants, and that discourse clearly informs his choice of the zucca as an occasion for exploring wild passions. The question which remains to be considered, however, is whether the problems of wild desire which Shelley stages in "The Zucca" could have been worked through successfully. As I noted earlier, the fact that Shelley abandoned the poem suggests that he could see no solution. But the kind of plant which Shelley chose—the gourd—might provide some clues as to the means by which he hoped to arrive at a solution, at least in his poem. In "The Witch of Atlas" (1820), Shelley describes a boat fashioned by Eros ("the first-born Love"), "like an horticultural adept," from a "gourd" which he has grown from "a strange seed" planted in the "mould" of "his mother's star," i.e. Venus:

> The plant grew strong and green – the snowy flower
>   Fell, and the long and gourd-like fruit began
> To turn the light and dew by inward power
>   To its own substance; woven tracery ran
> Of light firm texture, ribbed and branching, o'er
>   The solid ring, like a leaf's veined fan –
> Of which Love scooped this boat – and with soft motion
> Piloted it round the circumfluous ocean. (lines 298, 300, 301, 302, 305–312)

Shelley's photosynthetic imagery here anticipates the recovery of the ailing zucca. But could Shelley also have planned to describe that later plant, watered by the speaker's tears, producing a mysterious fruit capable of redeeming the problem of unfulfilled desire, perhaps as a boat in which the ailing speaker and his beloved might sail (as Shelley imagines in *Epipsychidion*) to some distant isle? Support for this possibility certainly comes from the final botanical work by Shelley which I want to consider here: the "Unfinished Drama," in which a lover's triangle and a woman weeping over a mysterious gourd again take centre stage.

As I have shown in Duffy (2015), the extant draft of Shelley's "Unfinished Drama" engages closely with the complex, interpersonal relations of Shelley's circle at Pisa—whom he may actually have intended to perform the finished play. Shelley's very heavily worked and incomplete draft of the longest of the three extant scenes focuses on a woman, based on Jane Williams, who has been abandoned by her unfaithful lover, based on Edward Trelawny, and who is loved, in turn, by a Youth, based on Shelley, whose affection she does not return. This woman, called Zelica in the draft, lives "near the springs of Indus," but both she and the Youth have been transported by mysterious and unrevealed means to a distant island, where the scene is set.[9] "Indian plants" feature alongside English flora in the garden of "The Sensitive-Plant" ("Part Second," line 30), and in Shelley's "Unfinished Drama," too, exotic and domestic flora coexist in a kind of universal garden reminiscent of Thornton's *Temple of Flora*.

When the Youth asks Zelica to explain how she reached the island, she describes a botanical dream which very much recalls, as Weinberg (305) also notes, the creation of the boat, by Love, in "The Witch of Atlas." A "spirit like a child" (reminiscent of Eros) appears "amid the plants of lower India / Which I had given shelter from the frost within my chamber," Zelica says, echoing the speaker's rescue of the zucca in that earlier poem (Fragment 3, lines 125–129, 138). This "spirit" places a seed in one of the pots, and when Zelica wakes from her dream, she finds that an actual plant has begun to grow. "I gave it with the other plants / Its share of water," she says:

> And day by day, green as a gourd in June,
> The plant grew fresh and strong, yet no one knew
> What kind it was . . . its stems and tendrils seemed,
> Like emerald knot-snakes mooned and diamonised,
> With azure mail and sheathes like woven silver;
> Its leaves were delicate and large, – You almost saw
> The pulses [   ]
> And all the sheathes that folded the dark buds
> Rose like the crest of the cobra di capel. (Fragment 3, lines 161–171)

In forging links between erotic longing (Zelica's, for her departed lover) and a mysterious plant in a specifically Indian context, Shelley might have been drawing on the comic verse narrative "The Enchanted Fruit; or, The Hindu Wife," by William Jones: a Linnean parody in which the attempt to procure a mysterious fruit leads to a princess and her five husbands (whom Jones characterizes, collectively, as a "five-mal'd single femal'd flow'r") being compelled to reveal their former erotic misdemeanours before being reconciled (13: 217). In addition, Zelica's quite specific description of her mysterious Indian gourd plant, with its serpentine characteristics and "golden . . . and purple velvet flower" (Fragment 3, lines 173, 177), may well derive from the illustration of "A Group of Stapelias" in Thornton's *Temple of Flora*, with which it seems to have much in common, granted that *stapelia* is an African plant, as Thornton makes clear (see Figure 6.2).[10]

Certainly in associating unfulfilled erotic desire with a gourd, Zelica's narrative would seem to return us to the scenario of "The Zucca"—and Shelley also reprises from that earlier work and from "The Magnetic lady to her patient" the image of an individual weeping over a plant, though in this case it is Zelica who weeps directly on her gourd, moved to tears by the songs which she sings of "forgotten" or "deserted" love. "I nursed the plant," she says, "and on the double flute"

Played to it on the sunny winter days
Soft melodies as sweet as April rain
On silent leaves, and sung those words in which
Hafiz makes Echo taunt the sleeping spring
Saying that young Desire, too long awake,
Sought her in the fountain

. . . .

And I would send tales of forgotten love
Late into the lone night, and sing wild songs
Whose sighing notes lie on the frozen heart,
Of maids deserted in the olden time,
And weep like a soft cloud in April's bosom
Upon the sleeping eyelids of the plant
So that perhaps it dreamed that spring was come
And grew to meet such harmonies and dews. (Fragment 3, lines 183–201)

Figure 6.2 "A Group of Stapelias" from Robert John Thornton, *Temple of Flora* (1812); reproduced by kind permission of The Getty Research Institute (Hathi Trust).

The whole scene is so reminiscent in form, and even in vocabulary, of "The Zucca" that it would seem clear that Shelley is making a second attempt here to resolve the problem of wild passions that he had established in "The Zucca" through the anthropomorphic and quasi-autobiographical figure of the wild plant.

Watered by the tears which Zelica sheds for her departed and unfaithful lover, her plant produces "a green and dewy embryon fruit"—*embryon* is a notably non-botanical adjective—before growing "out of the lattice . . . along the garden and across the lawn . . . to the margin of the glassy pool" (Fragment 3, lines 181, 207, 209, 213). At this point, Shelley's draft peters out. But it seems reasonable to assume that Shelley meant to imagine the creation of a boat from the gourd, as he had done in "The Witch of Atlas," by which means Zelica would be transported from her home to the island, where the difficulties of love could be resolved, as in Shakespeare's *Tempest*, one of the ur-texts for the "Unfinished Drama." But that Shelley abandoned this draft at a point structurally and conceptually equivalent to that at which he gave up on "The Zucca" suggests that he could find no satisfactory resolution to the problems of unfulfilled desire. Plants, it seems, could serve as anthropomorphic figures for wild desires but not, ultimately, as images for their resolution.

George notes that the erotic charge of Linnean taxonomy meant that "botany becomes a discourse of female sexuality in eighteenth-century literature" (2). In the botanical poetry written by Percy Shelley in the last six months of his life, however, the focus is much more on male sexuality: on the wild or transgressive desire for a woman, or the failure to return a woman's love, both of which scenarios had immediate analogues in Shelley's personal life at the time. For a final instance of this use of botanical imagery, we might point to what Kelley calls Shelley's "plant portrait" (*Clandestine Marriage* 228–229) of Rousseau in "The Triumph of Life" (1822). Rousseau, who admits to the narrator of the poem that "I was overcome / By my own heart alone," appears in the pseudo-phallic form of "an old root" (lines 240–241, 182). As this image makes abundantly clear, Shelley's botanical poetry tends fairly consistently to subsume the wildness of plants into a figure for exploring other, human kinds of wildness. In addition to its erotic charge, however, this figurative *dewilding*, I want to suggest, is something which Shelley's botanical poetry also inherits from contemporary botanical discourse.

As we have seen, in theory, practice, and reception, Linnean botanical discourse, during Shelley's lifetime, drew close and complex analogies, both literal and figurative, between plants and persons, between wild flowers and wild passions. Systematic botanical classification is, of course, almost by definition, a form of *dewilding*: the making epistemologically familiar of the ontologically distinct. But the paradox of the Linnean system, at least as understood and practised in late eighteenth- and early nineteenth-century England, is that the focus on sexuality meant that the epistemological *dewilding* of plants was often matched by an affective *rewilding* of the human. In other words, the anthropomorphic aspect of Linnean botany—or at the very least of its many literary and popular cultural appropriations—effectively superimposed the wildness of human emotions upon the wildness of plants, using the wildness of plants to represent the wildness of the human. And it is exactly this kind of superimposition which is visible in Percy Shelley's late botanical poetry and which makes those poems so different from much of his other writing about the natural world.

Notes

1 For Shelley's involvement with Jane Williams, and responses to it by critics and biographers, see Duffy (2015).
2 Recent studies of botany in the Romantic period include George (2017), Kelley (2012), and Kelley (2003).
3 Unless otherwise indicated, Shelley's work is quoted from Shelley (2016).
4 Percy Bysshe Shelley, "The Zucca," lines 76–77; quotations are from Shelley's draft in Bodleian MS. Shelley adds. e.17, as reproduced in Shelley (1986–2002) 12: 328–351.
5 Commentaries include: Frosch 381–383; Wasserman 189–190, 417; Weinberg 304–306.
6 For the biographical context of poem, see Duffy (2015) 623–625.
7 On *Mab*'s "revolutionary" form and content, see Duff, 54–114.
8 Important readings include: Caldwell, 221–252; Cronin, 199–215; King Hele, 209–210; Kelley (2012) 225–230; Maniquis, 143–150; O'Neill, 155–179; and Wasserman, 154–179. See also Shelley (2016) 779–782, and Shelley (1988) 3: 288–290.
9 Shelley, 'Unfinished Drama', Fragment 3, line 82. Quotations are from Shelley's draft in Bodleian MS. Shelley adds. e.18, as reproduced in Shelley (1986–2002) 19: 236–239, 246–283, 290–291, 294–295.
10 Shelley ordered Thornton's *The Philosophy of Medicine* in July 1812 (see *Letters* 1: 319). These lines are the only evidence that he might also have known *Temple of Flora*.

Works cited

Blank, G. Kim. *Wordsworth's Influence on Shelley*. Palgrave, 1988.
Caldwell, R. S. "'The Sensitive Plant' as Original Fantasy." *Studies in Romanticism*, vol. 15, 1976, pp. 221–252.
Cronin, Richard. "*The Sensitive Plant* and the Poetry of Irresponsibility." *The Neglected Shelley*, edited by Alan Weinberg and Timothy Webb. Ashgate, 2015, pp. 199–215.
Duff, David. *Romance and Revolution*. Cambridge UP, 1994.
Duffy, Cian. "Percy Shelley's 'Unfinished Drama' and the Problem of the Jane Williams Poems." *European Romantic Review*, vol. 26, no. 5, 2015, pp. 615–632.
Frosch, Thomas. "'More Than Ever Can Be Spoken': Unconscious Fantasy in Shelley's Jane Williams Poems." *Studies in Philology*, vol. 102, 2005, pp. 378–413.
Gaull, Marilyn. "Shelley's Sciences." *The Oxford Handbook of Percy Bysshe Shelley*, edited by Michael O'Neill, Anthony Howe, and Madeleine Callaghan. Oxford UP, 2013, pp. 577–593.
George, Sam. *Botany, Sexuality and Women's Writing, 1760–1830*. Oxford UP, 2017.
Goethe, Johann. *Versuch die Metamorphose der Pflanzen du Erklären*. Gotha, 1790.

Graglia, Giuseppe. *New Pocket Dictionary of the Italian and English Languages*. Livorno, 1818.

Hele, Desmond King. *Erasmus Darwin and the Romantic Poets*. Palgrave, 1986.

Heringman, Noah, ed. *Romantic Science*, SUNY P, 2003.

Jones, William. *The Works of Sir William Jones*. London, 1807, 13 vols.

Kelley, Theresa, *Clandestine Marriage: Botany and Romantic Culture*. Johns Hopkins UP, 2012.

———. "Romantic Exemplarity: Botany and 'Material' Culture." *Romantic Science*, edited by Noah Heringman, SUNY P, 2003, pp. 223–254.

Maniquis, Robert. "The Puzzling 'Mimosa': Sensitivity and Plant Symbols in Romanticism." *Studies in Romanticism*, vol. 8, 1969, pp. 129–155.

Morton, Timothy. "Nature and Culture." *The Cambridge Companion to Shelley*, edited by Timothy Morton, Cambridge UP, 2006, pp. 185–207.

———. *Shelley and the Revolution in Taste*. Cambridge UP, 1996.

Nicholson, William. *The British Encyclopedia*. London, 1809, 6 vols.

O'Neill, Michael. *Romanticism and the Self-Conscious Poem*. Clarendon, 1977.

Quinn, Mary. "Shelley's 'Verses on the Celandine': An Elegiac Parody of Wordsworth's Early Lyrics." *Keats-Shelley Journal*, vol. 36, 1987, pp. 88–109.

Shelley, Mary. *The Letters of Mary Wollstonecraft Shelley*, edited by Betty T. Bennett, Johns Hopkins UP, 1988, 3 vols.

Shelley, Percy Bysshe. *The Bodleian Shelley Manuscripts*, general editor Donald Reiman, Garland, 1986–2002, 23 vols.

———. *The Complete Poetry of Percy Bysshe Shelley*, general editor Donald Reiman and Neil Fraistat, Johns Hopkins UP 3 vols. to date, 2000 in progress, 3 vols. to date.

———. *The Letters of Percy Bysshe Shelley*, edited by Frederick L Jones, Clarendon, 1964, 2 vols.

———. *The Poems of Shelley*, general editor Kelvin Everest, Routledge, 1988 in progress, 4 vols. to date.

———. *Posthumous Poems of Percy Bysshe Shelley*, edited by Mary Shelley, London, 1824.

———. *Selected Poems and Prose*, edited by Jack Donovan and Cian Duffy, Penguin, 2016.

Thornton, Robert John. *The Philosophy of Medicine*. London, 1800.

———. *Temple of Flora*. London, 1807.

Wasserman, Earl. *Shelley: A Critical Reading*. Johns Hopkins UP, 1971.

Weinberg, Alan. "Shelley's Italian Verse Fragments: Exploring the Notebook Drafts." *The Neglected Shelley*, edited by Alan Weinberg and Timothy Webb, Ashgate, 2015, pp. 281–306.

Wordsworth, William. *The Major Works*, edited by Stephen Gill, Oxford World's Classics, 2008.

7 Wilding Europe and *Childe Harold's Pilgrimage*

Cassandra Falke

Wilding depends on two interrelated processes. First, there are the natural processes of earth's producing and reclaiming. Dandelions open overnight; kudzu creeps up walls. Tiny acts of wilding occur unseen among viruses and micro-organisms. Huge tectonic shifts urge mountains upwards and rumble down cities. A process-based understanding of wilding focussed on these natural phenomena has gradually replaced site-based understandings in the last fifteen years, especially in places where tracts of land that have never been developed become harder to find. Second, humans perform symbolic acts that delimit and animate our cultural image of wilderness. Some symbolic acts, like the designation of wilderness spaces, function via the law, while others accumulate power culturally. For example, to celebrate fifty years of the American Wilderness Act, the Smithsonian Museum of Natural History ran a "Wilderness Forever" photography contest and exhibited winning photos for ten months in 2014–15. Many of the photos, including the grand prize winner, capture mountain scenes reminiscent of nineteenth-century photographs taken by Carleton Watkins in Yosemite and William Henry Jackson in Yellowstone. Photographs by Watkins and Jackson played a crucial role in creating the national parks system and helped shape an image of wild land that remains powerful. The persistence of compositional elements—mountain in the background, water in the foreground, seasonal colour in between—in everything from contemporary photography contests to MacBook screen savers belies the power that the cultural circulation of images has over the ways wilderness is defined and experienced. Simultaneously, by buzzing, erupting, birthing, and breaking, natural forces "wild" spaces indiscriminately, whether they are designated to become a recognized wilderness or not.

Currently, these two processes are discussed separately. Generally speaking, ecologists, poets, and policy makers look to the natural processes that wild and rewild land, and historians and cultural critics historicize that looking, emphasizing the way it gives human shape to processes treated as purely natural. The first half of this chapter brings the processes of natural and human wilding together and discusses them phenomenologically, as parallel to the processes of earth and world discussed in Heidegger's "The

Origin of the Work of Art" from 1936. Building on his discussion of art's capacity to "open the world and keep it abidingly in force" (169), I suggest that art wilds as well as worlds, circumscribing possibilities for wildness but also keeping the wild in force. In the second half of the essay, I turn to the use of the term "wild" in Britain in the late romantic period. After looking at some of the ways the term circulated in quarterlies and poetry, I focus on *Childe Harold's Pilgrimage*, a poem that was significant for nineteenth-century perceptions of the wild.

If wilderness is, in part, a historically particular process of conceptualization, then entering the wilderness depends on a prior symbolic transformation of places into wilderness—a wilding of what was previously unknown or viewed as unwild—as cultivated or useful, merely threatening or not worth thinking about. *Childe Harold* played a crucial role in accomplishing that symbolic transformation. It helped re-romanticize the grand tour amidst the deluge of British tourists entering the continent after the Napoleonic Wars (Buzard 35, Elfenbein 32) and was instrumental in the founding of American national parks. In the words of James Buzard, grand tour takers learned from *Childe Harold* to "aspire to a condition of self-culture above and beyond the call of class" by carrying pocket-sized versions and Byron-inspired travel guides (38). Noah Comet makes a comparable argument about *Childe Harold's* influence on the perception of American wilderness, tracing the Washburn party's "Haroldism" in their 1870 report on Yellowstone (700). These nineteenth-century readers discovered a potential in *Childe Harold's Pilgrimage* that many critics of romanticism have overlooked: Byron offers a model of perceiving wilderness that frees readers to see nature's wilding as constantly occurring wherever they are. By attending to wilding processes at work in pathless woods, villages, and cities, as well as traditionally sublime mountainscapes, *Childe Harold* foregrounds the ubiquity of nature's wilderness-making work. To return to Heidegger's phrase, it keeps the wilderness "in force."

To wild and rewild

American poet Gary Snyder treats the wild like something that happens more than somewhere to go. His description of wilderness in a conversation with fellow poet Jim Harrison captures this idea of wilding beautifully. "Nature is what we are in," he states. The "'wild' really refers to a process—a process that has been going on for eons." And "'wilderness' is simply topos—it is areas where the process is dominant." He points out that "The wild works on all scales . . . The wild can be a wood lot. Even the vacant lot in the city can be wild" (11). According to this definition, nature is always wilding, so places that appear unwild are merely those areas in which human systems of control have, for now, gained the upper hand. The contemporary commitment to "re-wild" parts of Europe and the UK supports this conception. Although human actions and manners of habitation can interfere with and even permanently disable specific wilding processes, such as the proliferation of a particular species, the broad complex of wild forces that retake, repair, and rewild continue to operate.

In European and American contexts, a process-oriented understanding of wilderness has replaced an older understanding that defined wilderness in a human-centred and negative manner, based on the absence of "permanent or significant human habitation" (IUCN). Currently, the European Union coordinates the protection of wild land through Natura 2000, a network of wild spaces set aside for preserving threatened species (Natura 2000), and the Wild Europe Initiative (WEI), which makes policy recommendations for the practical achievement of the EU's wilderness promotion goals. The working definition of wilderness shared by Natura 2000 and the WEI is "an area governed by natural processes. It is composed of native habitats and species, and large enough for the effective ecological functioning of natural processes" (WEI). By emphasizing the role of natural processes in the production and maintenance of wilderness, this definition coordinates with Snyder's, which was formulated in an American context. However important it might be for conservation efforts to emphasize the uniqueness of "untouched" wilderness, a definition focused on human habitation (or its absence) reflects an anthropocentric bias. Crucially, wild forces exceed the spaces and concepts humans build. Natural wilding can decentre and humble human onlookers.

Of course, the concept of the wild, like all concepts, alters historically with the human forces of language and cultural production. Like other concepts, it strains for an adequacy it never achieves. Wilding is therefore a human process as well. In order to count as wilderness, an area must fulfil the conceptual requirements that are created and re-created by every generation. In his oft-cited essay, "The Trouble with Wilderness," William Cronon emphasizes how the Romantic conception of the sublime underpinned the celebration of mountainous landscapes as quintessentially wild. It shaped persistent cultural perceptions of the Arctic (Duffy), the tourist economy of the Alps (Ring), and the cultural demarcation of spaces in Britain where one might find the wild. A brief overview of titles from the period clarifies where these were: *Remarks upon North Wales* (Hutton, 1803), *A Tour to the Principal Scotch and English Lakes* (Denholm, 1804), *A Guide to the Lakes* (Wordsworth, 1810), *The Highlands and Western Isles of Scotland* (MacCulloch, 1824). The wilderness, in the discourses of tourism and exploration, was configured as geographically removed from the centres of industry and urban culture.

Wild places were feared or valued for multiple reasons, but the conception of wilderness as place-bound was widespread. In their "Historical Survey of European Wildness," Thomas Kirchhoff and Vera Vicenzotti suggest that understanding wilding as a process is fairly new historically, emerging only in the 1990s (456). They describe seven conceptions of wildness based on seven different worldviews that evolved between the enlightenment and the mid-twentieth century. It is worth glancing at these briefly for the systematic overview they provide of wilderness concepts operative in the 1810s and 1820s.

1 "Early enlightenment: wilderness as a place of unspoilt divine exaltation" (446)
2 "Liberalism: wilderness as an arena of war, an object of appropriation and a place of freedom" (447); linked to the picturesque via Gilpin's emphasis on ruggedness (448)
3 "Democratism: wilderness as a medium of self-experience for the autonomous subject"; "inner wilderness" is an "urge-driven, unregulated part of inner nature" (449); linked to the Kantian description of sublime experiences as a means of self-affirmation
4 "Transition towards Enlightenment critique: wilderness as a symbol of unalienated life and as a surrogate for virtuous community life"; exemplified by Rousseau (450)
5 "Early romanticism: wilderness as a place for roaming sentiment, original chaos and freedom from the constraints of civilization"; "true wilderness can never be reached" because romanticism anticipates an impossible earthly reconciliation of the darkly wild and peacefully utopic (451–452)
6 "Classic English conservativism: wilderness as a place of the natural sublime that fosters the drive for self-preservation and social pity"; exemplified by Burke (452)
7 "Classical German conservativism: wilderness as the sphere of urge-driven wildness, original power, authenticity and latent order"; urban liberal society considered the "symbolic or actual location of a detached, immoral and dissolute life," referred to as *Wildnis* (453).

Each of these positions remains discernible among romantic period writers, and they predominantly present wilderness as a sphere or a place. But alongside them another concept of the wild was emerging; the English romantics already had a sense of wildness as a process. Wordsworth's note on "Tintern Abbey" distils a recurring theme in his writing: "There is an active principle alive in all things: / In all things, in all natures, in the flowers / And in the trees, in every pebbly stone" (*Lyrical Ballads* 309). As Paul Fry notes, Wordsworth's vitalism extends at times to inorganic things (107), but the poet most often exemplifies thingly activity through flowers, trees, and stones. According to Jonathan Bate, the recognition of nature's active, living principle is part of "the legacy of Romanticism in our age of eco-crisis" ("Living with the Weather" 435); he cites the role of the weather in Byron's "Darkness" and Keats's "To Autumn." James McKusick notes that Coleridge, too, frequently describes things in nature behaving as not things but forces (*Green Writing* 43). Romantic poets not only observed wilding as an active principle; they directed readers' attention to wilding processes, keeping them "in force."

Wilding and worlding

Heidegger has been significant for a certain branch of romantic ecocriticism (Bate, *Song of the Earth*; Offord 42–46; Rigby 6–7, 84). Jonathan Bate, in particular, has articulated an ecopoetics oriented towards Heideggerian dwelling (*Song* 257–283). In opposition to this poetics, object-oriented ontologists (OOO) have argued that technological mediatization has destroyed the metaphors of horizonality which underlie the concepts of dwelling and worlding. There is too little space here to adequately counter Timothy Morton's dismissal of phenomenological dwelling and the related concept of lifeworlds (*Hyperobjects* 98–109). However, the reflections that follow on Heidegger's concept of "earth" and "world" do not rely on the "correlationist," human-centred (13) assumptions OOOérs critique. Rather than entering too deeply into the debate over how much Heidegger can be useful for environmental criticism, this essay has the more modest goal of borrowing Heidegger's concepts of earth and world from his pre-war essay "The Origin of the Work of Art" in order to clarify the two processes of wilding with which the essay began. Heidegger's notion of earthly force, as something from the earth but not divisible from the cultural processes of worlding, clarifies the ways that natural and cultural worlding relate to each other without falling back on the clear object-subject distinction that a nature / culture division seems to imply.

In "The Origin of the Work of Art," "earth" is aligned with things and events prior to human perception and conceptualization, and "world" is aligned with human perception and conceptualization, not only as individual acts, but also as products of culture, history, and language. Earth and world, like the two processes of wilding described above, are always in flux. Heidegger defines world as an "all-governing expanse" of relationality made manifest in "birth and death, disaster and blessing, victory and disgrace, endurance and decline" (167). A world is historically contingent and dynamic, "the ever-nonobjective to which we are subject" (170). Worlding, like wilding, is a process. "The *world worlds,* and is more fully in being than the tangible and perceptible realm in which we believe ourselves to be at home" (170). That tangible and perceptible realm, if we could imagine it as prior to our engagement with it as world, would be what Heidegger calls "earth." Earth includes the materiality of the planet. Heidegger associates earth with "Tree and grass, eagle and bull, snake and cricket" (168) as well as "the massiveness and heaviness of stone" (171) and "the quiet gift of ripening grain" (159). But earth is also more than material. It is "the clang of tone" (171) and the "emerging and rising . . . in all things" (168). Heidegger associates earth with a range of non-human actions that emanate mostly from the natural world but also from the "equipment" of shoes, musical instruments, building stone, and even language—things that, being made, conceal their own constructedness and become taken-for-granted parts of our world most of the time.

Earth, according to Heidegger, never reveals itself, but is only revealed through world. We cannot hear a cricket without having the concept of cricket, prior experience with the sound, and the capacity for recognizing sound patterns all at hand. And when we hear crickets, the sound comes laden with memories of summer nights and screen doors, or whatever historically contingent world first held the sound for us in relation to a whole. Crickets may be part of earth, but we know them as part of a world. Being earth, we never know them fully, but we operate daily in the moving balance of that not-knowing, since our experience with crickets in the world generally signals that we know enough about them. Earth, as Heidegger uses it, connotes those "dearest freshness deep down things," which, manifesting themselves in natural or everyday objects, deny their status as object and reveal themselves as events. This freshness, to complete the line from Hopkins, "lives" ("God's Grandeur" line 10).

The process of wilding contains both an earth-force and a world-force. The wildness of earth appears in loping animals, high mountains, and cracked seeds, and we often look at these and think of them as things, especially when their movement happens at a rate that humans cannot perceive, but every object in nature is equally and simultaneously an event and a centre around which events occur. Wilderness is the state these events tend towards. Small-scale and big-scale, these events are happening constantly all over the globe, but the recognition of their value varies enormously, as does the extent to which wilderness is allowed to flourish. Worlding affects how earth appears. The wilding process that is part of worlding is, like earth's wilding, dynamic, multifarious, and ongoing. It is full of systems of meaning making that sometimes conflict. Since world is not, in Heidegger's usage, "a merely imagined framework added by our representation to the sum of such given things" (170), the force of wilding involved with human worlds cannot be said to be a mere conceptual overlay that credits some landscapes, some magic measurement of biodiversity with the tag "wilderness." Things gain "remoteness and nearness" through the organization of worlds (170). This has to do primarily with the ways our directedness and attention (what phenomenologists call intentionality) are organized, but the question of proximity is not purely metaphorical when we consider remoteness and nearness to the wild. World has to do with our literal, physical placement in relation to ecosystems, to animal and human others, and to social resources, as well as with the organization of concepts about these relations in language. Humanities scholarship about wilderness often projects urban or suburban lifestyles onto its readers, obscuring the fact that 3.4 billion people in the world (including some humanities scholars) continue to live rural. One's relationship to wilderness varies according to the accessibility of places where large-scale natural wilding occurs, and according to the time one has to attend to wild processes. It is not only concepts that help turn our attention to wildness. Individuals may live in different worlds in relation to wilderness although they are, broadly speaking, within the same culture.

Nevertheless, "the world worlds" in a way that is culturally mediated, so the conceptualization of wilderness through images, experiences, storytelling, language, and law varies in culturally specific ways across history.

Art contributes to worlding. Heidegger says, specifically that a work of art "opens up the world and keeps it abidingly in force" (169). Art does not stop with revealing itself as stationary in the framed or raised object of a painting or temple but directs observers' attention back to their world. Although the world "is never an object that stands before us and can be seen" (170), a work of art can direct our attention to some taken-for-granted part of our world and help us to see just that part. Heidegger uses the example of Van Gogh's "A Pair of Shoes" (159–160), which for him gives visibility to the reliability of everyday things, to the woman who wears them, her work in the field, and her resilience. Also, a work of art can bring attention to processes of worlding, such as when a Greek temple provides the destination for a procession (167). As the example of the Greek temple signals, a work of art does not keep the world in force for every viewer, listener, or reader in the same way. No longer a destination of religious processions, the Parthenon now keeps in force, among other things, the idea of Athenian culture.

Returning to the example of wilderness photography with this function of art in mind, we can see that some art not only worlds but wilds. By portraying unpeopled biodiverse spaces and intimidating peaks as wild, photographs such as Rodney Lough's contest-winning portrayal of Wonder Lake in the Denali National Park (Smithsonian) reinforce a particular idea of wilderness. A person looking at a scene with all of the elements associated with wilderness will recognize the scene as wild. In order to see a wood lot as wild, as Snyder says is possible, our attention would need to be directed to the wildness of it; it would need to be culturally coded as wild. No single work of art changes the everyday perception of shoes or wood lots for everyone or even for one person for all time, but works of art contribute to the symbolic transformation of mountains into wild mountains, and wood lots into small spaces of wilderness instead of noise buffers in neighbourhoods or sources for pulpwood.

British romantic wilderness

Between 1801 and 1815, the Napoleonic Wars, industrialization, and the enclosure of common lands all worked to constrict the wilding of lands close to large populations of people in Europe and Britain. The usual story of romantic literature's response to these changes is that the wild moved out and became a place to go to, an experience for city dwellers to have away from home in landscapes that fit the period's increasingly well-defined expectations of sublimity and the picturesque. Through associations with geographical remoteness and sublime excess, wilderness was presented as something best pursued away from everyday life (Oerlemans 159; Poetzsch 6). But this is only half of the story. Alongside romantic wildernesses of lonely wanderers and great, dark peaks, there is also:

> a robust tradition of Romantic realism, especially a Romantic realism of the everyday—everyday suffering and injustice, everyday beauty and dignity, everyday lands and ecosystems and the human activities within them. This is a tradition in which wilderness or the wild is not sublimely looming out there, outside human dwelling and activity, but rather is found closer to home—in our lands, our communities, our cities, our bodies. (Cladis 836)

Romantic authors reckon with the force of the wilderness in these everyday realms as well as in landscapes that dramatize distance and inaccessibility.

Mark Cladis turns to Wordsworth for evidence of this tradition of the everyday wild in British Romanticism. He calls Wordsworth a "democratic-environmental poet" (839) because he pictures marginalized characters within natural environments where the wild is at work, contesting the idea that the wild is "a pristine sanctuary where . . . transcendent nature can be encountered without the contaminating taint of civilization" (Cronon 19). But the romantics' appreciation of natural beauty in everyday landscapes is hardly limited to Wordsworth. James McKusick reads "This Lime Tree Bower, My Prison" as contesting the idea that natural beauty necessitated remoteness. He suggests that for Coleridge, "the perception of 'Nature' need not be confined to settings of Big Wilderness; even the most humble garden plot, when seen through eyes awake to 'Love and Beauty,' can respond to our innate sense of wonder in the presence of things we have not created ourselves" (10). Clare challenges so completely the suggestion that wilderness must be remote that his acceptance into the romantic canon was predicated upon a critical shift in the understanding of romantic nature poetry. Work by Raymond Williams and John Barrell in the 1970s was crucial in this respect. The wilding tendencies of Keats's plants and bugs literally enclose and penetrate domestic spaces, running round thatch eves in a conspiracy to load and bless a house ("Ode to Autumn," line 4) and invading the stove through the cricket's song ("On the Grasshopper and Cricket," lines 11–12). As Stanley Plumly puts it, Keats "adjust[s] the sublime equation into an equity, something more low-ercase" (342).

Critical attention to wilding in spaces closer to home has led to a rethinking of the sublime. Noting that a long-term critical interest in sublime and picturesque elsewheres had obscured much of the period's sophisticated writing about the aesthetics of nature in everyday life, Markus Poetzsch traces the lineage of an alternative "quotidian sublime" (23–63) in the period. Plumly credits Keats with defining an "intimate" sublime (355), and Oerlemans articulates a concept of the "material sublime" based on Keats and Wordsworth (4, 24). The sublime, in these critical reevaluations, becomes a term for the aesthetic experience of wildness—wildness that operates in Coleridge's lime tree and Keats's vines, Clare's birds, and Wordsworth's daffodils; wildness that flourishes beyond human control, but that does not stop at the boundaries of human habitation. Beneath or alongside the romantic perception of wilderness as defined by Kirchhoff and Vicenzotti (an unreachable place of "roaming sentiment, original chaos and freedom from the constraints of civilization"), lurks an awareness of wild that transcends place—reachable but not containable, its chaos unfolding in thatch eves and lime tree bowers as well as mountainsides.

Even so, Byron's name remains absent in most discussions of this democratized, everyday wild. J. A. Hubbell, who has written the only book-length account of Byron's presentation of nature, sums up the poet's position in scholarly discussions of romanticism as follows: "Through the various movements of formalism, post-structuralism, and New Historicism[,] . . . Byron stood for a Romanticism of comparative social and cultural history, irony and comedy, cosmopolitanism, and satire" (2). Nature ended up on the other side of the polarizing counternarrative, with Wordsworth. Stephen Cheeke's *Byron and Place* deals very little with natural forces, focussing more on residues of history and Byron's rhetoric of authenticity. Kate Rigby's excellent *Topographies of the Sacred* examines romanticism as a reenchantment of nature. Her survey of English, German, and French authors leads her to conceive of romantic nature as "dynamic, self-generative, and animate unity-in-diversity, of which humans too are integrally a part" (12), a view perfectly compatible with a focus on wilding as a process, but Byron appears in the book only as Shelley's travelling companion to Mont Blanc (160–161). Tracing the "waves" of romantic eco-criticism from 1990 to 2018, Hubbell finds that Byron has yet to emerge as an important contributor to the romantic period discourse on nature.

In light of the critical neglect of Byron's writing about nature generally, it is no surprise that his references to the wild have gone unnoted. Contemporary criticism differs markedly in this respect from the criticism of Byron's contemporaries. Reviewers then did not hesitate to associate him with the wild. In his unsigned *Quarterly* review of *Childe Harold* III in 1816, Walter Scott praises Byron and Coleridge for "wild, unbridled and fiery" imaginations. He yokes "wild" together with "original" and "magnificent" (191) when praising Byron, while retaining the critique that both he and Coleridge sometimes wander too far "into the wild and mystic," leaving the reader "at a loss" to determine their meaning (204). The following year, John Wilson's review of the same text in *Blackwood's* also evokes "wild" as a key descriptive term. Wilson compares Byron's poetry to "a wild cry from another world," "his own wild world." Although these repeated allusions to the poetry's otherworldly quality seem to imply that wild realms are realms of exclusion, Wilson redirects his use of the term to indicate otherwise. Reading Byron, he says, "even those whose lives have had little experience of the wilder passions, for a moment feel that an unknown region of their own souls has been revealed to them" (112). This wild region of the soul is presented as parallel to a "sublime assemblage of woods, glens, and mountains—of lakes and rivers, cataracts and oceans." Byron's exploration of the "heart" of nature gracefully succeeds his exploration of "the heart of Man" (112). In Wilson's geography of souls, everyone has his or her own wild expanse, characterized by its resistance to being known and controlled. In writing about the wild as something within a human soul as well as in the "sublime assemblages" of nature, Byron characterizes wilding as a pervasive process occurring just under the surface of polite manners based on his perception of equally pervasive wilding forces in nature.

In the early nineteenth century, the term "wild" was not associated with distance to the same degree that the terms "sublime" and "picturesque" were, nor was it preserved for the observation of nature as something separate from human perceivers. Essay writers and poets recognized wildness as manifesting itself in humans, animals, and landforms equally. The term "wild" appears to mean "resistant to control," and is applied to all sorts of human and natural elements. Wilson's *Isle of Palms* is described as a "wild story" in an 1812 review in *Blackwood's* (376), as is Byron's *Mazeppa* in 1819 (429). Wilson uses the term liberally in his poetry; there is a "wild Indian glade" and "wild land" (113, 165), but looks and eyes are also wild (166, 160). This pattern is borne out if we look at one of Byron's favourite contemporary poets, George Crabbe. Having been born in 1754, Crabbe was a generation older than Byron, but his popularity reached its height around the time *Childe Harold* was published. In his 1812 collection, *Tales*, Crabbe refers to dreams and notions as wild, sneaking in as they do below the radar of consciousness. Crabbe's "wild" scenes include cut turf in one instance, and in another, dikes holding back the ocean, so the term does not indicate a complete inability to control, as much as the presence of something that resists control.

Byron in unwild Europe

Wildness is a powerful, if inconsistently deployed, concept in *Childe Harold's Pilgrimage*. Early in Canto Three, Byron states that he must think "less wildly," less "darkly" (lines 55–56, 20). What does it mean to think wildly? The context suggests that wild thinking relates to the powerful and destructive creativity thematized throughout *Childe Harold's Pilgrimage*, especially in Cantos Three and Four where Byron pictures Napoleon and Rousseau as key figures of both creativity and destruction (Hubbel 116–17 and Newey 155). But wild thinking is figured as much through natural forces as through Byron's consideration of other humans. These wild forces tend to be common but resistant to control—wind and night and storms and springs of water. Such forces refer to general powers of creativity and destruction that are not only a product of human world-making, but also and prior to that, ongoing parts of nature's wilding.

Byron's insistence that he "must think less wildly" immediately follows the famous stanza championing creativity as a source of vitality.

> 'Tis to create, and in creating live
> A being more intense that we endow
> With form our fancy, gaining as we give
> The life we image, even as I do now. (lines 46–50)

Wild thinking, these lines suggest, intensifies being itself through a creative act that simultaneously manifests and draws us into "The life we image." But in the midst of wild thinking, one's brain becomes "In its own eddy boiling and o'erwrought, / A whirling gulf of phantasy and fame" (lines 57–58). The "springs of life" become "poisoned" (line 60). Byron moves fluidly between figures of immense, obvious power (boiling whitewater) and small, hidden powers (poisoned springs). Perhaps because he often evokes natural forces as metaphors, it is easy to read right through the presence of nature in the poem and focus only on the ways springs or storms do conceptualizing work, but Byron's linking of wild thinking and nature's wilding reveals a profound receptivity to non-human forces at work in the earth (in Heidegger's sense of the term). The "being more intense," like the later mentioned "life intense," is life in which "not a beam, nor air, nor leaf is lost, / But hath a part of being" (III.838–840).

Byron's portrayals of nature in *Childe Harold* often pair creation and destruction in the setting of landscapes associated with sublimity, but he consistently foregrounds human finitude rather than taking refuge in a Kantian reconciliation of the sublime to human conception. This pattern recurs in one-line metaphors and extended verse sequences. After discussing madmen in Canto III, he writes, "Their breath is agitation, and their life / A storm whereon they ride, to sink at last" (lines 388–389). Storms, here and elsewhere in the poem, figure awe-inspiring power, but all that power is snuffed out by the end of the line. Just afterwards, he evokes "mountain-tops," the "loftiest peaks most wrapt in clouds and snow" (lines 397–398), and imagines the loneliness of ascending the summit, the classic sublime experience. The comparison of a "lofty mind" and sublime mountain top extends for four stanzas, and then a reference to the "chiefless castles" sitting near the summit shifts the focus to those who fought there now wrapped "in a bloody shroud" (lines 413–421). The power of humans' lofty thinking comes across clearly here, but again the encounter with sublime nature ends in a contemplation of death. In terms of the emotions evoked by the poem, the proximity of death may function as it frequently does in romantic evocations of the sublime; conjuring mortality is like looking at danger but not being in it. But in terms of what we might call the phenomenology of the sublime, the manner through which the subject relates to an encounter with great power, what Byron does is quite unique. The evocation of death denies the possibility of the perceiving subject's dominance. The conceptual subsumption of natural forces is associated with physical human dominance and shown to be illusory. Wild thinking encompasses the sublime's exposure to threat, but without seeking recourse in metacognitive reconciliation in the way that Kant describes. Death is not evoked as a threat that a brave mountaineer, chief, or poet evades, but as a fate we all meet regardless.

One of the Canto's most famous passages repeats this pattern of evoking the sublime and then switching to the contemplation of mortality.

> Could I embody and unbosom now
> That which is most within me, — could I wreak
> My thoughts upon expression, and thus throw
> Soul, heart, mind, passions, feelings, strong or weak,
> All that I would have sought, and all I seek,
> Bear, know, feel, and yet breathe — into *one* word,
> And that one word were Lightning, I would speak;
> But as it is, I live and die unheard,
> With a most voiceless thought, sheathing it as a sword. (III.905–913)

This passage is typically read, and rightly so, as a reflection on Byron's ambition for and frustration with self-expression. Sheila Emerson reads the stanza as "a carefully measured use of desperation" evoking the "highly specific impossibility" of a perfectly expressive word only to find it impossible (374–375). Beneath the metaphor, however, there is also the juxtaposition of one of nature's strongest forces and a resignation to death; the passage looks like an evocation of the sublime but actually rejects the elevation of human cognition which is the covert goal of sublime aesthetics. Jerome McGann reads this passage as evidence of Byron's "anti-natural" rejection of sentiment: "Byron's savage desire in this passage is therefore literally beyond nature . . . Byron's is the anti-nature that demands a morality beyond the order of moral virtue" (105). McGann aligns nature with the order of moral virtue that Byron seeks to get beyond. I would suggest that Byron finds nature—specifically nature in its pathless places and stormy moments, which is to say in those moments when its wildness is most apparent—already beyond an order of moral virtue. Lightning is already outside that order, a place where his language cannot go. In addition to emphasizing language's inadequacy to the expression of desire, the stanza acknowledges language's inadequacy to nature's force.

The stanza immediately preceding this describes natural forces that precede the self and grant the narrator the power of perception; he then turns back to them.

> Sky, mountains, river, winds, lake, lightings! ye!
> With night, and clouds, and thunder, and a soul
> To make these felt and feeling, well may be
> Things that have made me watchful. (III.896–899)

Watchfulness is a gift of "Sky, mountain, river, winds, lake" and "lightnings." The first interesting point here is that in spite of the rhetorical power of these stormy stanzas, the narrator is positioned as passive in relation to nature. Wild forces are not presented as subordinate to a perceiving subjectivity. Byron prioritizes the self being a material part of nature's strong reality over absorbing nature into the self. He asks, "Are not the mountains, waves, and skies, a part / Of me and of my soul, as I of them?" (lines 707–708). His being part of the strong reality of mountains, waves, and skies is the basis from which he begins. Whether mountains, waves, and skies are a part of him is questionable; whether he is part of them as natural material subjected to natural forces is known.

The second point is that, although the passage describes the narrator's own personal watchfulness, it presents "a soul / To make [night, clouds, thunder, sky, and wind] felt and feeling" as generic, not "my" soul, but "a" soul. That sort of soul is not limited to the speaker, but is listed alongside widely available experiences like the falling of night or the passing of a storm. It is also worth remembering that the "glorious night" that inspires the lightning stanzas precedes the description of Clarens, which Murray's *Handbook for Travellers in Switzerland and the Alps* describes as "a poor, dirty village far less attractive than many of its neighbours." The *Handbook* surmises that Clarens "probably owes its celebrity to its well-sounding name" (206). Percy Bysshe Shelley was similarly unimpressed, commenting in his "Letters from Geneva" that by the time he and Byron had arrived, "the spirit of the old times had deserted its once cherished habitation" (96). Apparently, it was not a destination that led every visitor to contemplate nature's wilding, but Byron does. Surrounded by glaciated peaks, he focusses instead on two phenomena that are common throughout Europe—thunderstorms and dawn. However dramatic dawn may be, it happens every day. By recognizing wilding as a natural force at work outside of areas more conventionally described as wildernesses, Byron makes the experience of wilderness more democratically available.

Skies and storms are there in poor, dirty villages, and they are there in cities. "Sky, mountains, river, winds, lake" and "lightnings" have made Byron watchful (III.896–899), but he then turns that watchfulness to grasshoppers (III.814) and leaves (III.839; IV.289), flowers on graves in Lisbon (I.272) and weeds in Rome (IV.231). He attends to small wilding forces as well as great ones. None of this is to say that the "hum of human cities" does not "torture" the "one who is not fit" to "stir and toil" in the "hot throng" of "mankind" (III.682–683; 653–657). As James Buzard notes, Byron spills plenty of ink deriding his fellow countrymen and women. "The most distant glimpse or aspect of them," the poet declaims, "poisoned the whole scene" (qtd. in Buzard 32). But by showing wild forces of perception to be generally available and wild forces of nature to be common, he opens a way for members of that hot throng in a city to become individual watchful souls.

Would Byron have called the forces of lightning and spring growth equally wild? Is this tamer wildness linked to wild thought? Possibly. In the first two cantos of *Childe Harold*, wildness tends to be associated with unthinking things: weeds, flowers, bells, locks of hair, and flocks of sheep (not individual sheep, who might think) roaming trackless mountains "all unseen" (I.132, 272, 507; II.222). When connected to animals, wildness is typically not linked to animals as individuals but their movements or sounds—the wild staring and plunging of a horse, the shrieking sew-mew (I.766, 121)—as if to dissociate it from the intentionality of either the animal or a human onlooker. These plungings and shriekings and locks of hair are all part of Heidegger's earth. Nature is wild when "nothing polished dares pollute her path" (II.329–330). Byron does not reject the idea that nature's wild is "polluted" by polished culture, but his wild is not a place or a feeling about a place. The worlding acts of the poem, especially in the first two cantos, idealize wild power in earthy acts and materials that concepts obviously fail to contain. Nature's wild is a persistent and uncontrollable force that expresses itself superfluously if viewed from an anthropomorphic perspective.

In Cantos III and IV, the force of the wild is increasingly associated with the poem's speaker, who desires to leave his "wild thinking" (III.55) and yet insists he cannot. He compares himself to a "wild-born falcon with clipt wing" (line 129). His "despair," like his thinking, is "wild" (line 141), and eventually his whole world (line 798). He refers to other wild people who exemplify wild thinking. Rousseau and his teeming love are both called "wild" (lines 725, 744), and Washington is reared in the "wild," his life a seed of freedom that "Earth" nourishes "Deep in the unpruned forest" (IV.860–862). Love and freedom, any form of passion, retain their association with "the world's wilderness" (IV.1079). However destructive the creation of wild thinking might be, it is contrasted with the "false creation" which "o'er-informs the pencil and the pen, / And overpowers the page where it would bloom again" (IV.1097–1098). Lightning-like, wild thinking is associated with creative force that leaves evidence of its reality but exceeds any attempt to conceptually contain it. Byron's presentation of wild thinking in *Childe Harold's* second half is predicated on his development, in the first half, of the wild as a force working unperceived and beyond human control all the time. The last mention of the wild in Byron's poem occurs in his apostrophe to the ocean, with its "wild waves" in the final stanzas (IV.1636). In the end, wild thinking is immersed again in nature's wilding and human thought approximates wildness only as a metaphor.

Byron contributes crucially to the "radical romanticism" that democratizes wilderness experiences (Cladis 837), accomplishing a cultural and symbolic wilding of Europe that becomes a model for perceiving the wild beyond Europe and away from the spaces presumptively designated as wild. Much as they may appear "beautiful and permanent" to Wordsworth, the "forms of nature" emerge in Byron's *Childe Harold* as not forms but forces, not permanent and not only beautiful but fierce and changeable: uncontrollable and demanding, small and overlooked (*Prose* I: 124). Situating human passion as one wild, unspeakable force among many, Byron contests the anthropomorphic view that humans can define or contain wilderness. He also highlights the pervasiveness and commonality of natural forces that resist our control. The sky is "wonderous strong" (III.861). "Glowing into day," the dawn compels our watchfulness (III.918). Attending to them is crucial to "the life intense" (III.838). And there they are, every day.

Works cited

Abrams, MH. "Structure and Style in the Greater Romantic Lyric." *From Sensibility to Romanticism*, edited by Gordon Haight and Harold Bloom. Yale UP, 1965, pp. 527–560.

Barrell, John. *The Dark Side of the Landscape: The Rural Poor in English Painting, 1730–1840.* Cambridge UP, 1980.

———. *The Idea of Landscape and the Sense of Place 1730–1840: An Approach to the Poetry of John Clare.* Cambridge UP, 1972.

Bate, Jonathan. "Living with the Weather." *Studies in Romanticism*, vol. 35, no. 3, 1996, pp. 431–447.

———. *Song of the Earth*. Picador, 2000.

Brady, Emily. *The Sublime in Modern Philosophy: Aesthetics, Ethics, and Nature.* Cambridge UP, 2013.

Byron, George Gordon, Lord. "Childe Harold's Pilgrimage." *The Complete Poetical Works.* Vol II, edited by Jerome McGann. Clarendon, 1980.

Cladis, Mark. "Radical Romanticism and Its Alternative Account of the Wild and Wilderness." *ISLE: Interdisciplinary Studies in Literature and Environment*, vol. 25, no. 4, Autumn 2018, pp. 835–857.

Comet, Noah. "Wild Childe: Byron and the Yellowstone Frontier." *European Romantic Review*, vol. 27, no. 6, 2016, pp. 699–718.

Crabbe, George. Tales. Project Gutenberg. 2004, http://www.gutenberg.org/files/5217/5217-h/5217-h.htm

Cronon, William. "The Trouble with Wilderness, or Getting Back to the Wrong Nature." *Environmental History*, vol. 1, 1996, pp. 7–28.

Denholm, James. *A Tour to the Principal Scotch and English Lakes.* Glasgow, 1804.

Duffy, Cian. *The Landscapes of the Sublime 1700–1830: Classic Ground.* Palgrave Macmillan, 2013.

Emerson, Sheila. "Byron's 'One Word': The Language of Self-Expression in 'Childe Harold' III." *Studies in Romanticism*, vol. 20, no. 3, 1981, pp. 363–382.

Fry, Paul H. "The Power of Things in Lyrical Ballads." *The Cambridge Companion to 'Lyrical Ballads,'* Edited by Sally Bushell, Cambridge UP, 2020, pp. 105–117.

Heidegger, Martin. "Building, Dwelling, Thinking." Translated by Albert Hofstadter. *Basic Writings.* Revised and expanded edition, edited by David Farrell Krell. HarperCollins, 1993, pp. 343–363.

———. "The Origin of the Work of Art." *Basic Writings.* Translated by Albert Hofstadter Revised and expanded edition, edited by David Farrell Krell. HarperCollins, 1993, pp. 139–212.

Hutton, William. *Remarks upon North Wales*. Knott, 1803.
IUCN: International Union for Conservation of Nature. "Category 1b Wilderness Area," https://www.iucn.org/theme/protected-areas/about/protected-area-categories/category-ib-wilderness-area.
Kant, Immanuel. *Critique of Judgement*, edited by Nicholas Walker and Translated by James Creed Meredith. Oxford UP, 2007.
Keats, John. *The Major Works*, edited by Elizabeth Cook, Oxford UP, 1990.
Kirchhoff, Thomas, and Vera Vicenzotti. "A Historical and Systematic Survey of European Perceptions of Wilderness." *Environmental Values*, vol. 23, no. 4, 2014, pp. 443–464.
Lough, Rodney Jr. "Smithsonian Gallery," https://www.rodneyloughjr.com/in-smithsonian
MacCulloch, John. *The Highlands and Western Isles of Scotland*, edited by Hurst Longman and Orme Rees. Brown, and Green, 1824.
McGann, Jerome. "Byron and the Anonymous Lyric." *Byron and Romanticism*, edited by James Soderholm. Cambridge UP, 2002, pp. 93–112.
McKusick, James C. *Green Writing: Romanticism and Ecology*. Palgrave Macmillan, 2000.
Murray, John (firm). *Handbook for Travellers in Switzerland and the Alps of the Savoy and Piedmont*. New ed., enlarged. John Murray, 1839.
Natura 2000. https://ec.europa.eu/environment/nature/natura2000/index_en.htm
Newey, Vincent. "Authoring the Self: Childe Harold III and IV." *Byron and the Limits of Fiction*, edited by Vincent Newey and Bernard Beatty. Liverpool UP, 1988, pp. 152–165.
Oerlemans, Onno. *Romanticism and the Materiality of Nature*. U of Toronto P, 2004.
Offord, Mark. *Wordsworth and the Art of Philosophic Travel*. Cambridge UP, 2016.
Phipps, Henry Constantine. *The English in Italy*. 2 vols. Saunders and Otley, 1825.
Plumly, Stanley. *Posthumous Keats*. Norton, 2008.
Poetzsch, Markus. "*Visionary Dreariness*": *Readings in Romanticism's Quotidian Sublime*. Routledge, 2006.
"Rev. of The Isle of Palms and Other Poems by John Wilson." *The Edinburgh Review, or Critical Journal*, vol. XIX, 1813, pp. 373–88.
Ring, Jim. *How the English Made the Alps*. Murray, 2000.
[Scott, Sir Walter]. "Rev. of Childe Harold's Pilgrimage III and Other Poems." *Quarterly Review*, vol. XVI, 1816, pp. 172–208.
Shelley, Percy Bysshe. *Essays, Letters from Abroad, Translations and Fragments*. Moxon, 1845.
Smithsonian. "Celebrating 50 Years of American Wilderness," http://www.wilderness50th.org/smithsonian.
Snyder, Gary and Jim Harrison. *The Etiquette of Freedom: Gary Snyder, Jim Harrison, and the Practice of the Wild*. Counterpoint, 2010.
United Nations. "68% of the World Population Projected to Live in Urban Areas by 2050, Says UN," 16 May 2018, https://www.un.org/development/desa/en/news/population/2018-revision-of-world-urbanizationprospects.html#:~:text=The%20global%20rural%20population%20is, by%20China%20(578%20million).
Wild Europe Initiative (WEI). "Wilderness and Large Natural Areas: Definitions," https://www.wildeurope.org/category/wildareas/definitions/
William Wordsworth, *Guide to the Lakes*, edited by Ernest de Sélincourt with a preface by Stephen Gill. Frances Lincoln, 2004.

———. *Lyrical Ballads and Other Poems, 1797–1800*, edited by James Butler and Karen Green. Cornell UP, 1993.
———. *The Prose of William Wordsworth*, edited by W.J.B. Owen and Jane Worthington Smyser, 3 vols. Oxford UP, 1974.
Williams, Raymond. *The Country and the City*. Oxford UP, 1975.
Wilson, John. *The Isle of Palms and Other Poems*. James Ballantyne, 1812.

8 Hölderlin, Heidegger, and hyperobjects

William Davis

In "The Question Concerning Technology" (1954), Martin Heidegger turns to the opening lines of Friedrich Hölderlin's late hymn "Patmos" (completed 1803, published 1808) in hopes of finding a way beyond the twentieth century's destructive relation to industrial technology:

Nah ist
Und schwer zu fassen der Gott.
Wo aber Gefahr ist, wächst
Das Rettende auch. (*Werke* 2.1:165, lines 1–4)

[Near is
And difficult to grasp, the God
But where danger threatens
That which saves from it also grows. (Hamburger 231)]

Typical of his engagements with Hölderlin, Heidegger does not attempt a literary reading of the poem, but rather deploys a noncontextualized snippet of verse as if it were the fragment of a sacred text crying out for exegesis in a time of modernist crisis.[1] He indeed cites only the third and fourth lines of the poem's opening, drawing from them a prophetic hope that where the "danger" (of a perverted relation to technology) threatens, grows likewise "that which saves." Hölderlin thus appears for Heidegger not merely as poet but as prophet, as "the poet who comes to us from the future" (Bernasconi 153), an attitude Heidegger expressed quite succinctly in the famous interview with *Der Spiegel* magazine in 1966 (not published until after his death in 1976) in which he insisted that we cannot view Hölderlin simply as one of many poets in the history of German literature but must reserve for him a category all his own: "For me Hölderlin is the poet who points into the future, who waits for the god, and who therefore cannot remain simply an object of Hölderlin Studies within literary-historical conceptions" ("Nur" 214; my trans.).

Hölderlin is the one who waits for the god, as Heidegger also waited, mid-twentieth century, agonizing over the state of the modern world, looking for salvation from the poet/prophet Hölderlin. Now, around a half a century after Heidegger's interview, we find ourselves (still waiting) within a discursive moment in which "the question concerning technology" may have become something more like "the question concerning the human." We have inherited the modernist crisis of industrial technology as the (often literally) burning calamity of the Anthropocene run amok. In this context of technology, the environment, and the human, I find it striking that Timothy Morton likewise invokes these opening lines of "Patmos," which he clearly takes directly from Heidegger, in his book *Hyperobjects* (2013). Although he by no means apotheosizes the poet as Heidegger does, Morton specifically links "that which saves" (he writes "saving power," following William Lovitt's 1977 translation of Heidegger) to his conception of the hyperobject: "Hyperobjects are a good candidate for what Heidegger calls 'the last God,' or what the poet Hölderlin calls 'the saving power' that grows alongside the dangerous power" (*Hyperobjects* 21).

Morton's citing of Heidegger's citing of Hölderlin raises the question of the relation between the poem and the contemporary discourse around technology and the human's place in the natural world, or rather, the posthuman's place in the *postnatural* world. This essay will be an attempt to take Hölderlin's poem seriously as we consider the sort of salvation Heidegger finds there, along with Morton's conception of it as a candidate for hyperobject status. Investigating these questions will lead us inevitably into the wilderness, into a place where, for Hölderlin, only a god can dwell. For Heidegger, wilderness stands as the element in nature that our technology cannot easily reduce to a "standing-reserve" of resources for our own consumption, which is also the realm of the "last god" on which he muses in the posthumously published *Contributions to Philosophy* (written 1936–1938) to which Morton alludes. This god for which Heidegger waits will not be "enframed" or tailored to fit any of our theologies. Morton, a supporter of "Ecology without Nature," rejects Romantic-Rousseauean notions such as "nature" and "wilderness" as inventions that have long outlived their usefulness: "to the extent that wilderness spaces . . . persist, we are still living, literally, within the Romantic period" (*Ecology* 114). He exchanges the *wild* for the *weird*, *wilderness* for *weirdness*, in this "time of the hyperobject," the time of things like plastic bags and global warming, things so "massively distributed in time and space relative to humans" (*Hyperobjects* 1) that they defy comprehension, and that thus tend to remind us of our "hypocrisy" and "lameness." As I understand them, for both Heidegger and Morton, the lines plucked from Hölderlin's poem appear to point to a form of salvation that rests upon a reorientation of our relation to objects, a call for us to resist the Romantic notion that we can become one with nature or make it our own. This notion of *thingness*, as it appears for both Heidegger and Morton as an untameable otherness, as a wilderness not reducible to human concerns, moves us closer to an understanding of the enigmatic opening of Hölderlin's poem, to the god we cannot grasp, however near.

As for the actual text of Hölderlin's "Patmos," both Heidegger and Morton give it pretty short shrift, which is unfortunate since the poem itself—beyond merely its third and fourth lines to which both allude—resonates clearly with ideas dear to both. Even taking a step back to consider the poem's first two lines—"Near is / And difficult to grasp, the God"—would go a long way towards connecting the poem to technology and hyperobjects, both of which concern ways in which we attempt to interact with, or "grasp," things that in their wildness and weirdness resist appropriation and annexation. This is a poem about waiting for a god that is difficult to grasp, a god that appears only to withdraw, much like those objects of object-oriented ontology (OOO) with which Morton is concerned.[2] To borrow a line from Heidegger's "Last God," "Patmos" is about "the advent as well as of the absconding of the gods" (*Contributions* 324).

By way of brief overview, this long hymn of 15 strophes and 225 lines, rife with complex syntax and allusions both classical and Biblical, opens with a lament for the absence of the divine, for "the isolation of the individual in godless times," as Friedrich Beisser puts it (*Werke* 2.2:789), which leads to a prayer for "innocent water" and "pinions" as compensation. As an apparent answer to this prayer, a divine being, a *Genius*, whisks him away eastwards to the dry and windy island of Patmos, the island of Saint John's Book of Revelation, much as we read in that book, "So he carried me away in the spirit into the wilderness" (KJV, Revelation 17:3). On this island, inhabited only by dry winds, or "voices of the hot noonday copse . . . where the sand falls, and the field's / Flat surface cracks" (Hamburger 233), the poet has his own vision that involves the death of Jesus, or his "absconding" (in Heidegger's sense) from the earthly realm. Though the disciples receive the Holy Spirit as compensation on the day of Pentecost, this appears to them as a mere shadowy substitute for the divine presence. The god is accessible now only in spirit. Lamenting this time in which "Too long, too long now / The honor of the heaven has been invisible" (Hamburger 242), the poet concludes that poetry ("German song") might constitute hope or renewal. Critics of the poem, those who actually try to read it as a literary text, concur that at its heart lies an elegiac longing for the divine in its absence, but they fail to note how strongly the desire these verses express is for a tangible connection to this elusive god of the wilderness.[3] Here the ideas of Heidegger and Morton can function as critical leverage in the service of reading Hölderlin's poem as a hymn on the *material* presence and absence of a god who forever evades our grasp. Although many elements of the poem lend themselves to such a comparative analysis, in the interest of space, I will limit myself chiefly to the first two strophes, which form the immediate context for the lines both Heidegger and Morton reference.

Near and difficult to grasp

> Wildness was a not-quite human force that addled and altered human and other bodies. It named an irreducibly strange dimension of matter, an out-side. (Jane Bennett, *Vibrant Matter*, p. 2)

If we examine the complete opening stanza of "Patmos," we see that it associates "danger" both with the god who is at once near *and* difficult to grasp, as well as with a location, a zone of danger that is a dark wilderness:

Nah ist
Und schwer zu fassen der Gott.
Wo aber Gefahr ist, wächst
Das Rettende auch.
Im Finstern wohnen
Die Adler und furchtlos gehn
Die Söhne der Alpen über den Abgrund
Auf leichtgebaueten Brücken.
Drum, da gehäuft sind rings
Die Gipfel der Zeit, und die Liebsten
Nah wohnen, ermattend auf
Getrenntesten Bergen,
So gieb unschuldig Wasser,
O Fittige gieb uns, treuesten Sinn
Hinüber zu gehen und wiederzukehren. (lines 1–15)

[Near is
And difficult to grasp, the God
But where danger threatens
That which saves from it also grows.
In gloomy places dwell
The eagles, and fearless over
The chasm walk the sons of the Alps
On bridges lightly built.
Therefore, since round about
Are heaped the summits of Time
And the most loved live near, growing faint
On mountains most separate,
Give us innocent water,
O pinions give us, with minds most faithful
To cross over and to return. (Hamburger trans. 231)]

The significance of a journey in darkness (*im Finstern*), to the "gloomy places" in Hamburger's translation, along with the danger there, becomes more evident if we attend to Hölderlin's conception of wilderness, a term that appears with some frequency in his writing at the turn of the nineteenth century. I find three distinct variations of his use of "wilderness" (*Wildniß* in his eighteenth-century orthography), which relate to each other through a dialectical tension. Wilderness functions geographically as a site of comfort and healing: "It is sweet to wander lost / in the sacred wilderness" ('Süß ists, zu irren / In heiliger Wildniß'), as we read in the poem fragment

"Tinian" (*Werke* 2.1:240). Related to this wilderness as *locus amoenus*, we find a temporal usage that suggests a time before the rise of civilization, a "glorious wilderness" that gave rise to a golden age now visible only as ruined fragments on the Greek landscape. On the opening page of Hölderlin's only novel, *Hyperion* (1799), we thus find the eponymous hero gazing down from the "heights of the Corinthian Isthmus" onto the ruins of the city of Corinth, which he imagines in its ancient glory as a happy medium between wilderness and civilization: "Between the glorious wilderness of Helicon and Parnassus where daybreak plays among a hundred snow-covered peaks and the paradisiacal plain of Sicyon, the shining gulf surged like a triumphant demigod toward the city of joy, youthful Corinth" (*Hyperion* 10).

In the fallen modern world, however, the progression of time itself has always already destroyed the "sacred" wilderness in which we might wish to wander, just as the "wilderness of Helicon and Parnassus" gave way to the civilization of ancient Corinth, and thus can persist only in poetic and mythical places such as "Tinian," a "distant, magical island" (Beissner, *Werke* 2.2:875). These two forms of mythical wilderness—geographic and temporal—comprise an interrelated dyad of a time or place that lies beyond the reach of contemporary human experience. Time destroys the mythical space that must then always be reestablished as that which is prior to civilization. The tension between the two gives rise to a third term that constitutes an expulsion into an ontological wilderness, or dark night of the soul, that strikes at the very heart of being—inevitable product of the modern condition. Exile from a temporal/geographic wilderness thus produces another that is psychic/spiritual, a wilderness within, which, though often represented through metaphors of dark and desert places, signifies the self-obsession and solipsism of the modern era. This is where Hölderlin's poem begins: among the dark places and chasms of the soul where only eagles dare.

Hölderlin's protagonist, Hyperion, fated to live at the end of the eighteenth century in a Greece under Ottoman rule, for example, deploys the term "wilderness" to express to his lover, Diotima, the conflicted states of the psychic condition: "So long as one melody still sounds for me, I do not shy from the deathly silence of the wilderness under the stars; so long as the sun only shines, and Diotima, there is no night for me" (90). Salvation grows where the danger lies, just as the darkness increases the returning power of the sun, or absence (even in "deathly silence") intensifies the longed-for return of the beloved, Diotima, who, following a well-worn Romantic trope, must herself soon die. In his protesting (perhaps a bit too much) against the persistence of night, Hyperion positions himself here as the eagle that is able to "dwell in gloomy places," or as one of the "sons of the Alps," who can walk "fearless over the chasm." The trick is to be able to "cross over and to return": *fort-da*.

Inherent in the modern fallen state that leaves us exiled to a "wilderness under the stars" and "growing faint / On mountains most separate" is a troubled relationship with the divine. We encounter god only as "difficult to grasp" (*schwer zu fassen*). On my reading, much turns in the poem on this concept of grasping. The German *fassen*, much like the English "grasp," has a clearly figurative as well as a literal meaning. We can thus read the poem's opening lines quite accurately as either "difficult to understand" or as "difficult to embrace," or more poetically, as both at once. "Patmos" is not merely about *comprehending*, but about *touching* the god—a fantasy of the substantial presence of the divine. Emmanuel Lévinas gets at this troubling doubled sense of the concept, when he speaks of knowledge as assimilation or appropriation. Our "discoveries," he argues, rather than uncovering something radically new, often result in more of the same: "even the most surprising discoveries end by being absorbed, comprehended, with all that there is of 'prehending' in 'comprehending'" (60). A god that we could easily grab on to, one easily "absorbed," "prehended," or "totalized" (to deploy another Lévinasian term), would not be a god at all.

For Hölderlin, the difficulty of grasping the god appears as a function of the god's very radical otherness that Lévinas finds so rare in our discoveries of the world around us. We are left waiting for this god who appears to possess a "wildness" that is "an irreducibly strange dimension of matter, an out-side," as Jane Bennett puts it in the quotation that heads this section, or an "audacious" and "remote" god, one not given to our comprehending or prehending. Hölderlin's god may be "near" but it is also "out-side" in Bennett's sense—"strange" to us. There lies the danger, but the salvation as well. For, as I hope will become apparent, following the logic of "Patmos," it is this very wildness as radical otherness of the god that is likewise our hope of salvation. Such a god can only be approached from Hyperion's "deathly silence of the wilderness under the stars."

In order to clarify the problem of god's precarious presence as it appears in the lines that concern Heidegger and Morton (more on them shortly), I turn to yet another passage from *Hyperion*, one that I believe presents us with a strategy for grasping the divine that, although seductive, ultimately fails in its effort to establish a viable connection to divine otherness. As Hyperion writes to his friend Bellarmin: "Often, lost in the wide blue, I look up at the ether and into the holy sea, and I feel as if a kindred spirit opened its arms to me, as if the pain of solitude dissolved into the life of the divinity. To be one with all–that is the life of the divinity, that is the heaven of man" (*Hyperion* 12). The open arms of nature signify an "all-embracing" god whose return Hyperion, like Hölderlin himself, continually anticipates: "Soon . . . there will be but one beauty; and mankind and nature will unite in one all-embracing divinity" (*Hyperion* 121). These visions of perceived oneness are of short duration, however, and Hyperion repeatedly finds himself "cast out of the garden" (13): "But a moment of reflection hurls me down . . . The world's eternal unity is gone; nature closes her arms and I stand like a stranger before her and do not comprehend her" (13). Nature, no longer "all-embracing" (*allumfassend*)—the German includes the verb *fassen*—but now "difficult to grasp," closes her arms, which amounts to an expulsion from the garden, a banishment into those dark and "gloomy places" where "the danger lies," into an ontological wilderness.

In the preface for an early draft of the novel Hölderlin designates this movement between feelings of unity with all and the despair of isolation as an "eccentric path" that we, like *Hyperion*, must walk—as if we were planets orbiting a sun to which we are at times close, but at other times hopelessly remote:

> We all must travel an eccentric path. There is no other possible way for us from childhood to adulthood. The blessed unity—Being, in the most fundamental sense of the word—is lost to us, and we had to lose it so that we would have to struggle to acquire it for ourselves. We tear ourselves away from the peaceful Εν και Παν [One and All] of the world in order to restore it within ourselves. We are part of fallen nature, and what, as we must believe, was once one, is now at war with itself. (*Werke* 3:236, my trans.)

Hölderlin's metaphor of the eccentric path serves as a commentary on the notion of an ontological wilderness I mention above. Just as Hyperion continually moves between the sun's light and a "deathly silence of the wilderness under the stars," we dwell within a "fallen nature," now "at war with itself"—signifying our pathetic moment in the dialectic. This is why the poet of "Patmos" prays for "innocent water," as well as for wings "to cross over and to return" (line 15)—"O pinions give us." We cannot make it on our own.

Only a god can save us

> Nur noch ein Gott kann uns retten. Uns bleibt die einzige Möglichkeit, im Denken und im Dichten eine Bereitschaft vorzubereiten für die Erscheinung des Gottes oder für die Abwesenheit des Gottes im Untergang; daß wir im Angesicht des abwesenden Gottes untergehen. (Heidegger, *Der Spiegel* interview, 209)
>
> [Only a god can save us. The only possibility for us is through thinking and poeticizing to prepare a readiness for the appearance of the god, or for the absence of the god in [our] downfall, as our downfall comes in view of the absent god. (Heidegger, *Der Spiegel* interview, 209, my trans.)]

When it comes to the "question concerning technology," the problem—or "danger" (*Gefahr*)—for Heidegger lies not in technology itself, but in the uses we make of it, in our interactions with it, which comes down to how we employ our machines and gadgets to *grasp* things. Put another way,

Heidegger is trying to tell us about proper forms of grasping (*fassen*)—as in comprehending, touching, and making use of objects around us—which again might make us wish that he had taken the first two lines of Hölderlin's poem into account since they point us towards the reason why Hölderlin's god, the "last god," in Heidegger's terms, is difficult to grasp, and *should be* difficult to grasp. It has to do with the containment, ordering, and marshalling of what would otherwise be wild, free-flowing, energy.

As is typical for Heidegger, understanding what he perceives as the "danger," as well as the potential salvation, requires entering into an adventure with German and Greek etymologies and wordplays. Rather than interacting with nature positively, which would be to engage it as a *Gegenstand* (object)—and here we have to understand that Heidegger means "object" in a good way—our technology tempts us to marshal the forces of nature as a "standing-reserve" that we can call upon whenever we feel the need, which is to turn a *Gegenstand* into a *Bestand*, a word that means "inventory" or "supply." Heidegger takes the prepositional prefix *gegen* (against) from the noun *Gegenstand* (typically translated as "object," but which we might render literally as something like *a standing-against*) as an indicator of something beyond the human that has not been fully appropriated by the human, something that stands over and against (*gegen*) the human. I read this prepositional force as not unlike Bennett's wildness of the "outside," mentioned above, as in something "irreducibly strange" that defies human appropriation. However, when we deploy technology to tame nature for our own purposes, we "order" (*bestellen*) it as well. The German verb *bestellen*, like English "to order," can mean both *to give orders to* and to *request something*. Heidegger cannot resist the wordplay: "Das so Bestellte hat seinen eigenen Stand. Wir nennen ihn den Bestand" (*Gesamtausgabe* 7:17). In Lovitt's translation: "Whatever is ordered about in this way has its own standing. We call it the standing-reserve" (*Question* 17).

Heidegger helps us out with an analogy of the Rhine river. Though it once flowed freely, perhaps romantically like a stream powering a "sawmill in a secluded valley of the Black Forest" (5), it has now been forced to flow through turbines that create a "standing-reserve" (*Bestand*) of electricity that we can call on at will (depending on how many people decide to charge their smartphones at any given moment, for example). Thus, the freely flowing river that was once a *Gegenstand* that remains *over and against* us, even as we interact with it—still *wild* in a strong sense of the word—becomes *Bestand*, an inventory of things that we can later order up according to our whims and desires. The "danger" for Heidegger lies not simply in the transformation of an otherwise wild world into a vast supply of resources ever at our disposal, "on tap," as Morton puts it (*Hyperobjects* 106). In our "ordering about" of nature through our use of technology we enter into a sort of Hegelian master-slave dialectic that threatens to turn the tables on us at any moment—mastered by our technology rather than masters of it—a nearly ubiquitous trope in our age of digital technology.

Our *un-wilding* of nature threatens to tame, commodify, and regiment us as well, which is where the real "danger" lies. We forget—to bring the discussion back to "Patmos"—that the Rhine is both close *and* difficult to grasp. Thus, in my view, the danger Heidegger outlines points directly to Hölderlin's use of "*fassen*": what we have here is a failure to grasp properly. By treating things as if we could order them about with our technology, we run the risk of losing that preposition *gegen* (against) in our relation to the world. Rather than bumping into a *Gegenstand*, into an object that might surprise or startle us in its unfolding, we run the risk of everywhere encountering only ourselves: "This illusion gives rise in turn to one final delusion: It seems as though man everywhere and always encounters only himself" (*Question* 27).

But just when things are getting dark and deluded, Heidegger calls Hölderlin to the rescue. Heidegger's first move with the two lines he adopts from Hölderlin is to interpret the word "save" (*retten*), which he takes from Hölderlin's *das Rettende* ("that which saves") as "to fetch something home into its essence [*Wesen*], in order to bring the essence for the first time into its genuine appearing" (28). A Heideggerian reading of the first strophe of "Patmos" thus suggests that it is the dark wilderness of the "gloomy places" that represents both the danger and the hope of salvation. Were it not "difficult" to understand and to hold the god, were the "all-embracing" of the divine never in question, the god itself might appear as too easy to get, as a "standing-reserve," as *Bestand*, rather than as an object that remains over and against us, untamed, a *Gegenstand*. Waiting for the god, longing with uncertainty for the return of the divine, positions the god always somewhere beyond the grasp of humans and their technologies. This is also where Heidegger's "last god" comes into play, "the god wholly other than past ones and especially other than the Christian one" (*Contributions* 319). Salvation, for both Hölderlin and Heidegger, lies in a god that refuses the logos of theology: "The last god has his own most unique uniqueness [*einzigste Einzigkeit*] and stands outside of the calculative determination expressed in the labels 'mono-theism,' 'pantheism,' and 'a-theism'" (*Contributions* 325–326; *Gesamtausgabe* 3:65, 411). Only a wild god can save us.

As noted above, it is because the god lies beyond our grasp that the poetic narrator asks in an apostrophe for "pinions" with which "to cross over and to return" (14–15). This crossing amounts to a journey into the wilderness, into the sun, and to the island of Patmos, but there grows also that which saves, the place where revelation is possible. Indeed, in strophe two we read that the poet's prayer is answered, after a fashion, with a granting of wings:

> So sprach ich, da entführte
> Mich schneller, denn ich vermuthet
> Und weit, wohin ich nimmer
> Zu kommen gedacht, ein Genius mich
> Vom eigenen Hauß'.

. . . nimmer kannt' ich die Länder;
Doch bald, in frischem Glänze,
Geheimnißvoll
Im goldenen Rauche, blühte
Schnellaufgewachsen,
Mit Schritten der Sonne,
Mit tausend Gipfeln duftend, so
Mir Asia auf. (lines 16–31)

[So I spoke, when more swiftly
Than ever I had expected,
And far as I never thought
should come, a Genius carried me
From my own house.

. . . no longer I knew those regions;
But soon, in a radiance fresh,
Mysteriously,
In the golden haze,
Quickly grown up,
With strides of the sun
And fragrant with a thousand peaks,
Now Asia burst into flower for me (Hamburger 231)]

Michael Hamburger's translation, "Carried me," is a bit tame, as we might well translate *entführte mich* as "kidnapped me." David Constantine gets a bit closer to the violence of the original with "took me away from home" (54). Being kidnapped and whisked away by a divine force is a "grasping" that leaves the poet in the position of object rather than subject. The divine does not arrive "all-embracing" (*allumfassend*), nor does the poetic narrator find himself "one with all that lives" as does Hyperion in his ecstatic moments. In "Patmos" the god's messenger appears with a startling abruptness that shocks and surprises: "more swiftly than ever I had expected," and "far, where I never thought to go" (my more literal translation). "Far" (*weit*) stands in clear contrast to "near" (*nah*) of the opening line. Who knows where the god, in its "most unique uniqueness," might take us? We are not in Swabia anymore.

In an attempt to express a complex line of thought briefly, we might say that Heidegger reads the extracted lines—"But where danger threatens / That which saves from it also grows"—as a prophetic suggestion that the very human impulses and abilities that allow us to create complex technologies with which we can order nature about, an activity he calls an "enframing" (*Gestell*), might also be put in the service of revealing (*entbergen*) the essential qualities of things. Drawing on Plato, Heidegger uses the Greek *poiesis* to designate this positive and creative activity of "bringing forth" the essence of things in such a way that truth in a strong sense (beyond mere facticity) is the result. This sort of truth also deserves a Greek name—*aletheia*. All of this is etymologically hidden in "technology" itself, which derives from a Greek term for art, *techné*. Since the human capacity for technology includes already within it our capacity for art, where the danger lies grows also that which saves.

Extending these Heideggerian ideas beyond the two lines he extracts as prophecy allows us to consider that it is this very difficulty of grasping—understanding, holding onto—the divine that pushes us towards the creative form of revealing, *poiesis*, that the philosopher has in mind. It is only because the god is "difficult to grasp" within the wilderness of "gloomy places" that the poet is driven to a prayer for wings, a prayer that results in the journey towards the sun and Patmos, towards epiphany. The revelation of "Patmos" thus transpires only as the result of an effort to reach out and touch a god that does not like to be touched, a god that "stands outside of the calculative determination" as does Heidegger's "last god." The god of "Patmos," resisting both "prehending" and "comprehending," will not be enframed or turned into a standing-reserve. It is paradoxically the banishment to the ontological wilderness with which the poem begins, the place where danger lies, that acts as the prerequisite for revelation.

Hyperobjects

> What kind of weirdness are we talking about? Weird weirdness.
> —Timothy Morton (*Dark Ecology* 7)

As we have seen, Morton's object-oriented ontology (OOO), or weird realism, perceives a gap or rift between how things are and how they appear to us: "the gap between phenomenon and thing yawns open, disturbing my sense of presence and being in the world" (*Hyperobjects* 12). The hyperobject, in its blunt vastness and ubiquity, "massively distributed in time and space relative to humans" (*Hyperobjects* 1), makes us hyper-aware both of this gap and of our "lameness" in the face of it: "Hyperobjects have already ushered in a new human phase of hypocrisy, weakness, and lameness" (2). Regarding Morton's assertion that "hyperobjects are a good candidate for what Heidegger calls 'the last God,' or what the poet Hölderlin calls 'the saving power' that grows alongside the dangerous power" (*Hyperobjects* 21), I suggest that if Heidegger's "last God"—the one that will not be enframed by "calculative determination"—might serve as a hyperobject, rather than turning to the "saving power," we would do better to turn our attention to Hölderlin's god as well, the one that is both near and difficult to grasp.

To get a somewhat better idea of the sort of (hyper)objects Morton has in mind, it is helpful to start with a version of *thingness* he opposes, particularly since this phenomenological foil arises out of the Romantic discourse of which Hölderlin was a part—we have already seen an example of it in *Hyperion*. I have elsewhere used the term "vibrant objects" to refer to a phantasmatic understanding of natural objects that Romantic poets and philosophers deploy as a means of overcoming the Subject-Object split of Kantian dualism (Davis 61–62).[4] On the philosophical side of such thinking we find, for example, Friedrich Schelling, Hölderlin's one-time friend and roommate, who argued, in his *Naturphilosophie* (Philosophy of Nature) of the 1790s, for the notion that nature, including both animate and inanimate life, is vibrantly shot through with one, absolute, life force, which he sometimes Neoplatonically called "the World Soul" (*die Weltseele*). On this view, all *things* are interconnected with the same vibrant energy, guaranteeing that subjects are not hopelessly cut off from Objects. "The Spirit (*Geist*) moves within All," as Schelling puts it (*Durchs Herz* 28). It is this conception of nature that allows Hyperion to meld with the sky and the sea, to be one with all. In the wake of Kant and Fichte and what we might call the "subjectivity wars" of the 1790s—we subjects over here, standing in uncertain relation to objects over there—Romantic philosophers and poets found ways to bring the world back together by inventing a variation of Rousseauean nature that grants fluid vibrancy to things such that they comprise matter and spirit at once, making them far easier to grasp.

Morton argues that, rather than relying on fantasies such as *vibrant objects* that function as reconcilers of the subject-object scission, we would do better to stop thinking in terms of such subjects and objects altogether. We are, he likes to point out, "always inside an object" (*Hyperobjects* 17). The material world clings to us. We cannot climb outside in order to become independent subjects, gazing back on the place from which we came. As I noted above, Morton thus eschews concepts like nature and wilderness as unhelpful remnants of our Romantic past. In their place he prefers "weird," a word he employs with great frequency, as anyone who has read him knows. We get a good sense of the reasons for his predilection for the "weird" in a section from *Dark Ecology* (2016) in which we find him, Heidegger-like, seeking after etymologies. "Weird," he tells us, has old Germanic roots in the "Fates or Norns," a causal sense that links it with *worth* as a verb, as in *to become* (also related to modern German *werden*), thus: "In the term weird there flickers a dark pathway between causality and the aesthetic dimension, between doing and appearing, a pathway that dominant Western philosophy has blocked and suppressed" (*Dark* 5). In other words, and this is central to our concerns here, "the dark pathway"—much like Hyperion's "dark wilderness under the stars"—suggests that the weird lies in the "*rift* (*chōrismos*) between *essence* and *appearance*" ("Object-Oriented" 209; emphasis in original). The divine (the sun for Hyperion), or the last god, always evades our grasp. Or, as Morton puts it in a succinct definition of his philosophy: "OOO . . . thinks of everything as a weird entity withdrawn from access, yet somehow manifest" (208). What, we might ask, is "somehow manifest" yet "withdrawn from access" if not another way of expressing "near and difficult to grasp"?

Following Morton's logic, Hyperion's Romantic fantasy of oneness-with-all thus forms part of the problem rather than the solution. The desire to meld with vibrant objects appears as one of those chapters of Western philosophy that wants to occlude the "dark pathway" between essence and appearance, to repress the existence of such gaps through a compensatory fantasy of blending into oneness. In other words, Hyperion, when "lost in the wide blue" and feeling as if "a kindred spirit opened its arms to me" (*Hyperion* 12), is not weird enough for Morton. Objects (*Gegenstände* in Heidegger's sense), as they stand over and against us, are weirder than spirit (*Geist*). Hyperion only begins to feel truly weird when fate casts him to the far side of the eccentric path, into the ontological wilderness:

> I reflect and find myself as I was before, alone, with all the pains of mortality; and the asylum of my heart, the world's eternal unity, is gone; nature closes her arms and I stand like a stranger before her and do not comprehend her . . . I am now isolated in the beautiful world, cast out of the garden of nature, where I bloomed, and am drying up under the midday sun. (*Hyperion* 13)

God as hyperobject withdraws. Here again we find the sentiment behind the poet's prayer: "Give us innocent water, / Opinions give us, with minds most faithful / To cross over and to return" (Hamburger 231). Waiting for the god who is always near but impossible to grasp, we are dying of thirst, like Tantalus.

In Morton's terms, waiting for the god in this manner signifies our "lameness," a millennial pejorative that for him has "a very specific resonance" (*Hyperobjects* 2). We experience a sense of "lameness" when confronting the "rift" that continually appears between our limited experience of the hyperobject and the object itself, or in Hölderlin's case, between us and our god. "The fundamental reason for lameness has to do with a special property of any given entity that is particularly visible in the case of hyperobjects. An object fails to coincide with its appearance-for another object, no matter how accurate that appearance-for" (196). For Morton, though we do not know much, we know at least that such vastness extends ontologically far beyond what we can see or comprehend, beyond what we can grasp. Wherever the essence of the thing may lie, because we are lame, we would need "pinions" to get there.

This reading of the prayer for wings thus speaks also to the apparent paradox of the god at once near yet difficult to grasp. "Nearness," which Morton also calls *viscosity*, is one of the hyperobject's primary characteristics: "I do not access hyperobjects across a distance, through some transparent medium. Hyperobjects are here, right here in my social and experiential space" (27). He also points out, however, that "while hyperobjects are near, they are also very uncanny" (28). They do not conform readily to our conceptions of them. Something may be right in our face, yet we cannot comprehend, prehend, or apprehend it. Yet, just as Heidegger finds "the saving" already growing within the danger, as the *techné* within technology, so Morton sees salvation in the lameness of the rift. These reflections indeed bring him back to Hölderlin:

> When hyperobjects are fully exposed to human being, the power and freedom of the Romantic sublime inverts itself into contemporary lameness. It is not an unpleasant reversal, unless you really need to be on top all the time. Indeed, the "saving power" that Hölderlin speaks of is truly the saving lameness of the Rift: the way in which all things, humans no exception, are hobbled from within by an ontological gap. (194)

As my reading of the opening strophes of "Patmos" suggests, in his poem Hölderlin appears to write against the grain of the Romantic sublime of Hyperion's "one with all that lives." The poetic narrator of "Patmos" finds himself ontologically at that end of the eccentric path that is farthest from the sun, as if just having experienced Hyperion's "moment of reflection" that "hurls [him] down" (*Hyperion* 13). The god is there, like the sun that remains in view, but like the hyperobject, cannot be grasped. Dying of thirst and aware of his lameness, our poet is clearly "hobbled from within by an ontological gap," in Morton's terms, or cast into what I have called Hölderlin's "ontological wilderness." As the poem puts it, the poet is one of those "growing faint / On mountains most separate" (Hamburger 232). In this sense I agree with Morton that there is something of the "saving lameness of the Rift" in Hölderlin's poem. It is the recognition of his own lameness, as well as of the chasm between himself and his god, that drives the poetic narrator to the prayer that instantiates the epiphany of the journey to Patmos. We can thus read those opening lines as an expression of ontological distance between the poet and his "absconding" god, a god that, again like the hyperobject—in its very proximity—tends to render the gaps between presence and absence painfully evident.

I conclude by returning to Heidegger's "Last God," a concept that connects all three of our authors. On my reading of Heidegger's dense and difficult text, the notion of the god yet to arrive works as a counterpart to Christian eschatology, particularly fitting since Hölderlin's "Patmos" is directly involved with the Book of Revelation. Heidegger does not cite Hölderlin's poem here, however, at least not overtly, yet there are striking similarities to its opening strophe. For example, Heidegger's last god is one of simultaneous "remoteness" and "nearness," which in Morton's terms also defines the hyperobject, as we have seen: "somehow manifest" yet "withdrawn from access" ("Object-Oriented" 208). In Heidegger's words: "The extreme remoteness of the last god in the refusal is a peculiar nearness" (326). "Near is / and difficult to grasp, the God."

The advent of Heidegger's last god is meant to herald a renewed relationship to being that is authentically connected to truth as *aletheia*, as an uncovering, to "the essence of truth of unconcealedness" (Risser 133). This ability to reveal requires what Heidegger calls a "refusal" or "denial" (*Verweigerung*) of being, which we might read as a refusal to "totalize" or "prehend" in Lévinas's sense, a refusal to turn what was weird or wild into a standing-reserve, *Gegenstand* into *Bestand*. The capacity to deny oneself in favour of revealing the truth of otherness, in turn, depends upon an experience of privation as loneliness, of the god's absence, which is to say time spent in Hölderin's "gloomy places, in "the deathly silence of the wilderness under the stars." Only those with the ability "to cross over and to return" (as Hölderlin's poem puts it) can apprehend the last god. Once again, we find that it is this very exile into the ontological wilderness that constitutes the danger as well as that which saves. Heidegger here calls this danger the "plight (*Not*) of the abandonment by being," which is also the "extreme remoteness of the last god" (*Contributions* 326). Those who experience it comprise their own lonely wilderness as they "are always removed from one another, peaks of the most separated mountains" (326). From "Patmos" again:

> . . . round about
> Are heaped the summits of Time
> And the most loved live near, growing faint
> On mountains most separate. (Hamburger 231)

In this (perhaps unconscious) near quotation of "Patmos," Heidegger again designates Hölderlin as "the poet who points into the future, who waits for the god" ("Nur" 214). It is only in such wild places as the "mountains most separate," or on Patmos, "where the sand falls, and the field's / Flat surface cracks," that we can hope to encounter the last god. This is where we find the poet, still waiting.

Notes

1 Critics have often accused Heidegger of using Hölderlin to suit his own ends, rather than attempting to read the poet carefully. See Andrzej Warminski: "Heidegger has no trouble whatsoever quoting Hölderlin out of context-indeed, some would say that his whole project of interpreting Hölderlin rests on arbitrarily ripping lines out of context and making them mean something other than what they mean 'in context'" ("Monstrous" 197). Thomas Ryan likewise suggests that, when writing about Hölderlin, Heidegger's literary criticism is "dubious at best" and that he relies on "primal sympathy" rather than on "textual grounding" (37).

2 Proponents of *Object-Oriented Ontology* (OOO), which Graham Harman and Timothy Morton also call "weird realism" ("Well-Wrought" 188, "Object-Oriented" 207), and which is part of larger philosophical movement generally called "speculative realism," frequently refer to the "withdrawal" of objects from perception. As Harman puts it: "Real objects withdraw from all human

access and even from causal interaction with each other" ("Well-Wrought" 188). Likewise, Morton: "OOO is a form of weird realism, in which objects have an essence that is profoundly withdrawn. Even when objects appear to touch one another physically, they are withdrawn from one another ontologically" ("Object-Oriented" 207).

3 David Constantine gets to the heart of the interpretive consensus when he states that, "the theme of the poem is absence and separation, the longing for the recovery of the fulfilled life, and how one should live in the meantime" (258). I do not disagree but wish to stress the significance of material/bodily "absence and separation" for the poem.

4 For the term "vibrant object," I lean on Jane Bennett's notions of "thing-power" and "vibrant materiality"—the ability objects have "to act as quasi agents or forces with trajectories, propensities, or tendencies of their own" (vii).

Works cited

Bennet, Jane. *Vibrant Matter: A Political Ecology of Things.* Duke UP, 2010.

Bernasconi, Robert. "Poets as Prophets and as Painters: Heidegger's Turn to Language and the Hölderlinian Turn in Context." *Heidegger and Language*, edited by Jeffrey Powell. Indiana UP, 2013, pp. 146–162.

Constantine, David. *Hölderlin.* Clarendon P, 1988.

Davis, William S. *Romanticism, Hellenism, and the Philosophy of Nature.* Palgrave Macmillan, 2018.

Harman, Graham. "Technology, Objects and Things in Heidegger." *Cambridge Journal of Economics*, vol. 34, 2010, pp. 17–25.

———. "The Well-Wrought Broken Hammer: Object-Oriented Literary Criticism." *New Literary History*, vol. 43, no. 2, 2012, pp. 183–203.

Heidegger, Martin. *Contributions to Philosophy (of the Event).* Translated by Richard Rojcewicz and Daniela Vallega-Neu. Indiana UP, 2012.

———. *Gesamtausgabe.* Klostermann, 2000.

———. "Nur ein Gott kann uns retten. Interview with *der Spiegel*, Sept 23, 1966." *Der Spiegel*, vol. 23, 1976, pp. 193–219.

———. *The Question Concerning Technology and Other Essays.* Translated by William Lovitt. Garland Publishing, 1977.

Hölderlin, Friedrich. *Hyperion or the Hermit in Greece.* Translated by Ross Benjamin. Archipelago Books, 2008.

———. *Sämtliche Werke.* Ed. Friedrich Beissner. Grosse Stuttgarter Ausgabe. Cotta; W. Kohlhammer, 1946–1985.

———. *Selected Poems.* Translated by David Constantine. Bloodaxe Books, 1996.

———. *Selected Poems and Fragments.* Translated by Michael Hamburger. Penguin, 1998.

Lévinas, Emmanuel, and Philippe Nemo. *Ethics and Infinity.* Translated by Richard A. Cohen. Duquesne, 1985.

Morton, Timothy. *Dark Ecology: For a Logic of Future Coexistence.* Columbia UP, 2016.

———. *Ecology without Nature.* Harvard UP, 2007.

———. *Hyperobjects: Philosophy and Ecology after the End of the World.* U of Minnesota P, 2013.

———. "An Object-Oriented Defense of Poetry." *New Literary History*, vol. 43, no. 2, 2012, pp. 205–224.

Risser, J. *Heidegger toward the Turn: Essays on the Work of the 1930s*. State U of New York P, 1999.

Ryan, Lawrence. *Hölderlin's Silence*. Peter Lang, 1988.

Schelling, Friedrich. *Durchs Herz der Erde: Sämtliche deutschen Gedichte und poetischen Übersetzungen*. Verlag Ulrich Keicher, 1998.

———. *Friedrich Hölderlin*. Twayne, 1984.

Warminski, Andrzej. "Monstrous History: Heidegger Reading Hölderlin." *Yale French Studies*, no. 77, 1990, pp. 193–209.

———. "'Patmos': The Senses of Interpretation." *Modern Language Notes*, vol. 91, no. 3, 1976, pp. 478–500.

9 "Almost Wild"

Jane Austen's dirtiest of heroines

Colin Carman

What wild imaginations one forms, where dear self is concerned!

—Jane Austen, *Persuasion* (215)

It was dirty, indeed, but what did that signify?

—Jane Austen, *Northanger Abbey* (106)

At the heart of Jane Austen's sensitivity to nature is the close proximity between her major heroines and wild nature, specifically grounds and dirt. In what follows, I confine my attention to a trio of nature-lovers—Elizabeth Bennet, Catherine Morland, and Fanny Price—who are the wildest and dirtiest of her major protagonists. Getting down-and-dirty is a feminine badge of honour but only when an alertness towards wildness is balanced with a striving for propriety and rational self-rule. These objectives constitute what Austen exalts as "well-grounded reason" in *Mansfield Park* (331) and, in *Emma*, "forbearance, patience, self-controul" (262). If the process of socialization requires cleanliness—looking, smelling, and sounding clean (without the use of "dirty" words)—regression into one's original state involves a return to the wild. In an effort to "green" Austen's fiction, and to open her major novels to an ecocritical perspective, I will show how *Pride and Prejudice, Northanger Abbey*, and *Mansfield Park* work in concert with what Olli Lagerspetz, author of *A Philosophy of Dirt* (2018), calls the privileged status of *Homo sapiens* as uniquely sanitary, or "*Homo sordidus*—not merely the rational animal but also the dirty (and clean) animal, the animal whose life is in many ways regulated by ideas of the dirty and the clean" (12). Dirt is a metonym for the wildness of nature and the pre-social nature of the human, a species defined not solely by its reason but by its aversion to dirt in both material and moralistic terms. Given that the twin goals of every Austen heroine are the embodiment of ladylike behaviour (or, feminine propriety) and the acquisition (through marriage) of a gentleman's stately home, the discourse of wildness is an important means through which women such as Bennet, Morland, and Price distinguish themselves from other ladies on the marriage market. Both "propriety" and "property" are derived from the Old

French *propre*, which means "clean" and "pristine." Only when the affinity for grounds and dirt is repressed can these ladies truly lead, and only by controlling their "wild fit[s] of folly" (to borrow another phrase from *Mansfield Park*) can they catch the eye of a single man in possession of a good fortune, so long as "good fortune" is defined by the propriety and property deserving of their love and devotion (324).

Austen's fiction is seldom read in relationship to the environment. The natural world is usually treated as background when it is anything but. For Helena Kelly, author of *Jane Austen: The Secret Radical* (2017), the very name "Austen" is usually understood as a synonym for "pretty young women, big houses, *Pride and Prejudice*—demure dramas in drawing rooms" (13), while for Beth Lau (2015), Austen's "novels of manners may not appear to participate in her contemporaries' love of the nonhuman world" though many of her key love scenes "take place out of doors" (21). Lau is right that the novelist is rarely grouped with a contemporary like John Keats who expressed his enthusiasm for what he called "wild nature" in a variety of ways (322). More critical thought should be directed towards why Austen is generally excluded from her nature-loving contemporaries, or, at the very least, how an eco-sensitivity underlies her championing of cultivated women who have seemingly broken free of their natural roots and joined the upper echelon of gentry women who deign to get dirty. With the exception of Jonathan Bate who, in 1999, argued that the demarcation between culture and nature in Austen's fiction is a blurry one at best, and that her "ideal England" is defined by her characters' sense of "environmental belonging" (545), the settled view amongst scholars is that Austen was a social realist whose primary interest lies not in nature but in culture. Yet this is the same genteel author who, in her letters, admits: "Pictures of perfection as you know make me sick & wicked" (335). A few crucial scenes in which dirt figures prominently in Austen's fiction allow us to think about the imperfect, the unladylike, and the anti-social role that wild nature plays in her novels.

Recent scholarship, by Faflak (2014) and Singer (2018), has illuminated the dimensions of Austen and affect, even "wild" affect, and the ways that emotive states must be checked in order for a heroine's *Bildung* to proceed (Singer 96).[1] I want to cross-wire these two schools of thought (environmental and affect studies) to generate a new way of approaching three of her female leads who, when feeling and looking wild, engage with the soil in important ways. A concession at the start: just because a writer arranges her characters *al fresco*, and sets some of her love scenes outdoors, does not mean she is ecologically alert. One would be hard pressed to describe Austen as a nature-writer. That said, there are a few key scenes that commentators reference when considering her representation of the natural world. There are the strawberry beds, "rambling and irregular," at Donwell Abbey in *Emma* (312). Mrs. Elton prattles on about Mr. Knightley's strawberries as the "best fruit in England . . . [the] finest beds and finest sorts" (313). This is more nationalistic than ecocentric, for Donwell Abbey is a microcosm for

England itself (viewed reverentially by Emma as "untainted in blood and understanding") (313). Donwell Abbey and Pemberley (Mr. Darcy's country estate) are both "pictures of perfection," as much as Austen resists that phrase, emblematizing the ideal English estate and family seat. For Noah Heringman (2004), what spurs Elizabeth to let down her defences and give in to Mr. Darcy's romantic advances is the sight of his grounds, which he calls a "stylized forum for this healthy disorder: the woods are wild and 'luxuriant,' but their occasional openings offer startling glimpses of orderly abundance" (241). In the role of observer, Elizabeth is bewildered by the sight of her future home as that place where nature and culture are harmoniously blended: "She had never seen a place for which nature had done more, or where natural beauty had been so little counteracted by an awkward taste" (259). Austenites are at their most cultish when it comes to Mr. Darcy and the character traits that magnetize Elizabeth to him (e.g. his wit, his wealth, his good looks and taciturnity), but rarely do they mention that he has the good sense not to improve, or re-design, the grounds around Pemberley. This has more to do with good taste—versus "awkward taste"—than the novel's hero being some kind of tree-hugging steward of the land, and Elizabeth's admiration for her future husband's good taste spans from his management of Pemberley's grounds to his choice of furniture and artwork inside. "Every disposition of the ground was good," Austen writes, before breaking the elegant tone of her palatial pictorial with her use of the adjectival "wild." While Elizabeth and her aunt are marvelling at a portrait on the wall, Mr. Darcy's housekeeper approaches and explains that the painting's subject, having "gone into the army," has "turned out very wild" (260). Here is the perfect contrast, framed in aesthetic terms, between the good taste of Darcy and the ungentlemanly wildness of the novel's military man, George Wickham. Outside, Elizabeth sees the semi-wild grounds laid out like landscape art while, indoors, Darcy's servant operates like a museum tour-guide, putting into words what a portrait of Wickham cannot.

To contrast the wild id of Wickham, there is a similar but ironized usage of the adjective "wild"—"almost wild"—earlier in *Pride and Prejudice* (72). "Wild" first appears in Chapter 8, the dining room scene, which establishes the major virtues of her most popular heroine, Elizabeth Bennet: her autonomy, her intelligence, and her devotion to family. The chapter begins with the mean-spirited gossip of Miss Bingley and Mrs. Hurst, two ladies who perceive themselves as the Bennet sisters' social superiors. Jane has come to their rented property as a guest (and potential bride for Bingley's brother, Charles) but fallen ill, which sets her younger sister's rescue effort in motion. They are dumbfounded by the sight of this soiled stranger who, despite her mother's objections, has marched "three miles so early in the day" and "in such dirty weather" to nurse her sister back to health (70). More shocking is that Elizabeth travels freely without a male chaperon, which was the custom of the day. Along the way, she muddies her petticoat and reddens her complexion through the physical exertion required to make the trek from her home, Longbourn, to Netherfield. Such an act endears Austen's heroine to the reader, whereas the physical consequences of this sororal sign of solidarity ("hair so untidy, so blowsy . . . her ancles [sic] in dirt," as Bingley and Hurst comment) offend the other women's snobbish sensibilities (72–73). Elizabeth operates as an extension of wild nature, or that which cannot be controlled or kept outside of the domestic structures a fine house such as Netherfield symbolizes. Having made it through the wilderness, Elizabeth makes a dramatic entrance by dragging the dirt inside. By doing so, she identifies with her rural environs, which is the point of origin for any ecocentric sensibility. Consequently, Bingley and Hurst look on with disdain because the socially inferior Elizabeth has brought the outdoors in and scandalized their sense of propriety.

The relationship between wildness and dirt is easy to overlook. Elizabeth would not be one of Austen's most beloved figures if she lacked the ability to regulate her wild emotions. In the fictive context of *Pride and Prejudice*, Austen provides the reader with both indirect dialogue ("She had no conversation, no stile, no taste, no beauty," relays the narrator) *and* something far nastier, in the words of Louisa Hurst: "I shall never forget her appearance this morning. She really looked almost wild" (72). *Almost* wild: the near-wildness, or quasi-wildness of her heroine, is central to the following argument wherein I connect the English landscape (its dirt, mud, rock, and sand) with more characterological notions of wildness, or emotional excess.[2] Mrs. Hurst and Miss Bingley are gossiping, of course, or, as the colloquialism goes, dishing the dirt. Austen would have known the close association between gossip and dirt since she was well-versed in Samuel Richardson and his bestseller *Pamela; or, Virtue Rewarded* (1740). One passage in *Pamela* condenses many of the connotations that the word "dirt" carried in eighteenth-century culture. Richardson writes: "It might be well enough, if you were descended of a Family of Yesterday, or but a Remove or two from the Dirt you seem so fond of" (257). Such moralizing is channelled by Elizabeth's greedy mother—the butt of many jokes in *Pride and Prejudice*—in a similar caveat that precedes her three-mile walk. "How can you be so silly," Mrs. Bennet cries, "as to think of such a thing, in all this dirt! You will not be fit to be seen when you get there" (69). This approbation has two meanings: first, Mrs. Bennet asserts that such a walk would be a walk of shame versus a walk of pride and, second, she rightly predicts the derision with which their social superiors, Hurst and Bingley, will see her daughter. Elizabeth holds her ground, and, in one of the first bold demonstrations of her fixity of purpose, puts her sister's health before the pressures of propriety.

Neither Elizabeth nor her mother is present to hear the women's insults but the latter character echoes her daughter's critics when she cries, in the following chapter, "Lizzy . . . remember where you are, and do not run on in the wild manner that you are suffered to do at home" (79). A young woman hoping to enter the marriage market must not "run on," much less *walk on* in the dirt and rain but repress her innate instincts and, preferably, stay indoors. Like the petticoat, or underskirt itself, unladylike wildness must be moderated. The eminent Austen scholar Claudia Johnson (1988) claims that the dirty petticoat is not about hygiene but an allusion to Elizabeth Hamilton's *Memoirs of Modern Philosophers* (1800) and its depiction of the hysteric, petticoat-soiling feminist. Elizabeth is thereby joining a particular political tradition and relying on a costume to further establish the egalitarian ground on which she demands to be seen.

Johnson is minimizing matters, however, when she declaims that Austen's image is "not cleanliness, but rather the unseemliness of female 'energy'" (23). She overlooks the fact that dirt can never be divorced from its myriad connotations, ranging from the political ("getting dirt on one's opponent") and commercial ("dirty money," "dirty work," "dirt cheap") to the pornographic ("dirty books," "dirty old man," and "talking dirty") and scatological ("dog dirt"). As Sabine Schülting reminds us, the term "dirt" is fundamentally amorphous, signifying any kind of abjectness. So many things are dirty and *get* dirty in Austen's writings. The wallpaper, for example, inside the Crown Inn is described by Emma Woodhouse as "dreadfully dirty," and as yellow and worn-out as the interior's wainscot (237). Earlier in *Emma*, her hypochondriacal father transfers his anxiety to his future son-in-law when he thanks Knightley for his visit, saying: "But you must have found it very damp and dirty. I wish you may not catch cold" (58). Knightley boasts that he can withstand the dirt, mud, and dampness of the English countryside, with this riposte: "Dirty, sir! Look at my shoes! Not a speck on them" (58). The humour here is subtle: Knightley's character is spotless, inside and out, for he himself embodies the "self-controul" that he expressly admires in the character of Emma's rival: Jane Fairfax (262). Knightley knows that by presenting himself as spotless, he is conveying his own respect for an ancestral house like Hartfield (the seat of the Woodhouse family line) and its reputation for propriety. His protestation, "Dirty, sir!", articulates Austen's own view that a truly proper house requires the cleanest of occupants. In *Mansfield Park*, Fanny's younger brothers, a pair of "rosy-faced" schoolboys, are "ragged and dirty" (382), but they are merely taking after their father, a dockyard worker, who swears and drinks. "He was dirty and gross," Austen observes, and the unsurprising outcropping of his property: an "abode of noise, disorder, and impropriety" (390). Such a description throws the linkage between dirtiness and impropriety into stark relief. If Portsmouth was a dirty place, so, too, was Bath, the inland spa-city where Austen, a one-time resident, never felt clean.[3] In a letter to Cassandra, from 1808, she describes the Bath Hotel as the most "uncomfortable quarters—very dirty, very noisy, and very ill-provided" (125). Stockings can be dirty, as in Elizabeth's "dirty stockings," but this is the unfortunate result, Austen informs her reader only lines later, of "dirty weather" (70). Here, "dirty" and "rainy" are synonymous, as in "It was too dirty for Mrs Allen to accompany her husband to the Pump-room" (*Northanger* 100).

In terms of the outdoors, or the elements, seasons are dirty. There is the "gloom and dirt of a November day" in *Mansfield Park* (220) but also the "dirty month of February" (378). Winter is the dirtiest season of all. According to Mrs. Elton in *Emma*: "In summer there is dust, and in winter there is dirt" (311). This declaration is not far off from the author's own view. Writing from London, Austen reports to Cassandra that their brother's London residence is all "dirt and confusion, but in a very promising way" (212). The identical phrase occurs in a debate over English landscape design wherein Mary Crawford complains that improvements being made to her uncle's cottage in Twickenham (a village outside London) had disrupted her ability to move around freely *and* cleanly. "For three months we were all dirt and confusion," says Mary, "without a gravel walk to step on, or a bench fit for use" (85). The implication here is that shrubbery was a way to absorb the muddiness of a damp, even wet, ecosystem, and that gravel walks were a man-made way for people to walk, and sit, without getting dirty.[4]

While the discourse of dirt is right on the surface in a scene such as this one, the rhetoric of wildness is far subtler. Nineteenth-century conceptions of wildness were not only inextricable from looking "dirty" but also from the pressures on ladies to repress, as Mrs. Bennet puts it, their "wild manner" (79). If not, they run the risk of facing social opprobrium. There is a certain disgracefulness associated with dirt in Austen's writings; another letter to Cassandra, from 1798, includes a weather-report (not uncommon) in her correspondences: "There has been a great deal of rain here for this last fortnight . . . indeed we found the roads all the way from Staines most disgracefully dirty" (17). How dirt could operate as a social "disgrace" is harder to understand than how, more concretely, a dirty road is an impediment to progress, closing traffic in both directions. Dirt is the roadblock to advancement, vehicular or cultural; dirt connotes perplexity, regression, and backwardness. Dirt presents an obstacle in other ways, too. "The equation was," in the words of Lagerspetz, and still is: "*Cleanliness = Self-discipline = Progress.* Cleanliness was enlightenment, responsibility, health, modesty and light. Dirt was superstition, sloth, sickness, and darkness" (153). Lagerspetz's formulation sounds Foucauldian, which it is, but it also springs from psychoanalytical theory. Classical psychoanalysis tells us that dirt signifies a lack of sexual hygiene and, in some extreme cases, an obsessional object that precludes sexual satisfaction.

These cultural connotations of dirt have already been explored by Kathy Justice Gentile whose recent essay "Dirty Girls, Dirty Books, and the Breakdown of Boundaries in Jane Austen's Fiction" is a superb contribution to literary dirt studies. Gentile focusses on the ways in which the materiality of dirt menaces the dominant social order, specifically the "boundaries and restrictive definitions of women's roles in late eighteenth-century British society" (58). The aim of her essay, however, is more aligned with feminist criticism rather than eco-criticism and singles out Catherine Morland and Elizabeth Bennet as the dirtiest of Austen's heroines; the latter figure, according to Gentile, threatens to pollute the family line so aggressively guarded by Lady Catherine de Bourgh. I do not wish to duplicate Gentile's (dirty) laundry list of references to dirt in Austen's fiction. Dirt, mud, rock, and other geological features are good, earth-bound places to start when conceiving of a green Austen, but dirt is no simple matter when considered in relation to wild nature *and* wild manners as they are embodied by her fictional figures.

Consider the fact that *Northanger Abbey* charts the ways in which Catherine Morland learns to better manage her wild imagination and her original love of dirt. "She was moreover noisy and wild," she writes, "hated confinement and cleanliness" (39). Similar to Elizabeth's pride in dirtying her petticoat, Catherine's "love of dirt" presages the heroine's later attachment to gothic fiction, satirized by Austen as dirty books that lead women imaginatively astray. Published posthumously in 1817, the opening chapter—a *Bildungsroman* in miniature form—announces that the quasi-heroine's "love of dirt gave way to an inclination for finery" and that Morland "grew clean as she grew smart" (39). Surely this anticipates Lagerspetz's equation, "*Cleanliness = Self-discipline = Progress*," and societal pressure to grow up and abandon the wild child inside. Becoming a respectable member of English society demands that Catherine leave her wild ways behind, and while she can wash the dirt away, a "love of dirt" is a dimension of inward wildness that is far deeper and more indelible. On the dermal level, Austen writes, "her complexion improved; her features were softened," which merely externalizes a deeper development, which only her mother recognizes as Catherine's "personal improvement" (39). The objective of such personal self-improvement is propriety, or the good manners that a lady must perform not in the wild but inside a proper house like Pemberley and Donwell Abbey.

Later in the narrative, while seated together in a theatre box and dressed in finery, Catherine and friend Isabella Thorpe share a feeling of wild excitement. This scene is further testament to the fact that, in Austen, wildness can be checked but never entirely repressed. Fantasizing about the sight of Henry Tilney, whom Catherine says she does not see within the theatre, Isabella jealously likens her friend to an animal: "You really have done your hair in a more heavenly style than ever: you mischievous *creature*, do you want to attract every body? . . . Oh! what would I not give to see him! I really am quite *wild* with impatience" (my emphasis 89). Catherine succeeds in mollifying her friend's erotic excitability, or what Keats once termed "wild ecstasy" (461) but, in a later interaction with Tilney, she fails to control herself. Catherine appears to have absorbed Isabella's wild affect once she and Henry come face-to-face: "Not with such calmness was he answered by the latter: 'Oh! Mr Tilney, I have been quite wild to speak to you, and make my apologies. You must have thought me so rude; but indeed it was not my own fault" (109). This is a regrettable outburst on Catherine's part and the equivalent of donning a dirty petticoat inside Netherfield. Her shame is compounded when the clothes-conscious Mrs. Allen, ignoring the social cue, thinks only of her own finery: "'My dear, you tumble my gown,' was Mrs. Allen's reply" (109). In Catherine's case, she takes two steps forward but one step back; while she succeeded in hiding her anger towards Isabella, her wildness proves more difficult to conceal.

This is social awkwardness at its worst—a "cringe-worthy moment," as we say in today's parlance—and by failing to take Catherine's side, Mrs. Allen only amplifies her anxiety that she has been discourteous in a social setting. There is a defensiveness in the claim, "indeed it was not my own fault," which is a direct echo of Austen's letter to Cassandra, referenced at the outset: "If I *am* a wild Beast . . . It is not my own fault" (212–213). The difference is that, set inside a theatre, the awkwardness that circulates triangularly amongst Catherine, Tilney, and Allen is meant to be funny in a theatrical sense. The reader, joining Allen and Tilney, laughs at the heroine's expense. In effect, Catherine Morland has regressed back to the land, back to wild nature, and *back* to the tomboy status and gender equivocation that precede sexuation. The candour inherent in her apologia, "I really am quite *wild*," endears her to Henry Tilney and thereby lays the groundwork for their eventual nuptials and the novel's happy ending. Her love of Henry supplants her girlish love of dirt, only fulfilling the punning predestination buried in her surname: Catherine is *more* than the wildland, and more than dirt; propriety demands that she learn to regulate her "wild imaginations" and to settle upon a gentleman who matches Catherine's own class privilege. Marriage, then, is the reward for being on one's best behaviour, which is to say, reigning in one's innermost wildness and love of dirt.

Finally, there is Fanny Price, the earthiest of Austen's heroines who has attracted environmentally oriented attention for two important reasons. First, Fanny is closer to animals than the other protagonists, which Barbara Seeber has keenly explicated in her *Jane Austen and Animals*. Second, *Mansfield Park* is an account of the popularity of improvement, or landscape design, during Austen's day, and Fanny is the nearest thing to a naturalist insofar the natural world is a source of inspiration that lets her imagination run wild. She repeatedly rhapsodizes about what she calls the "sublimity of Nature" (135), which Austen contrasts with the vain and selfish Mary Crawford who, being a Londoner, feels out of place in the countryside. Mary lacks what Austen calls "Fanny's delicacy of taste, of mind, of feeling," for the feltness of Fanny's delicacy is inextricably tied to "nature": the English countryside, with its "roads," "harvest," "cottages," "cattle," "children," and, yes, its "soil," as in "the difference of soil" (106). In her dialogue with Mary, Fanny's gleeful observations of the natural world are studded with exclamation points, as in "I am so glad to see the evergreens thrive!" and "how astonishing a variety of nature!" (224). Fanny explains: "You will think me rhapsodizing; but when I am out of doors, especially when I am sitting out of doors, I am very apt to get into this sort of wondering strain" (224). As in *Pride and Prejudice*, and the scene in which

Elizabeth finds herself under the gaze of English ladies, Fanny's sense of self is refracted through other people's opinions of her (i.e. "you will think me"), which predetermines the way in which a cosmopolitan lady like Mary will judge her enthusiasm for wild nature—the same "rapture" that she shares with Edmund earlier (135).

What further heightens her wonder is the fact that the English countryside was undergoing drastic change during the period, through such cultural guidelines as parliamentary enclosure and improvement. The sixth chapter of *Mansfield Park* is a detailed account of the high-priced efforts, amongst England's landed gentry, to "improve" the landscape, for better or worse. Mr. Rushworth embodies this trend and, when he first visits Mansfield Park, he has come directly from a friend's property where the grounds were being "laid out by an improver," which left his "head full of the subject" (80). The chapter, too, is full of the subject, for it stages the debate over whether nature should be left wild and free or be improved by landscape architects. Austen even names names, via Miss Bertram's recommendation to Rushworth: "Your best friend upon such an occasion," she says, "would be Mr. Repton, I imagine" (81). Humphry Repton (1752–1818) was the leading improver in Austen's day, travelling some 600 miles monthly and undertaking more than 400 commissions before an accident in 1813 (the year of *Pride and Prejudice*'s publication) confined him to a bath-chair.[5] The irony, as Kim Wilson puts it, that "the ultimate result was just as much an artificial production as what it replaced," was not lost on Austen (28). Repton was fashionable but also controversial in his day, for he helped to shift the garden, and the practice of landscape gardening, from the exclusive domain of aristocrats to the middle classes. The poet William Cowper, whose poem *The Task* (1786) Austen greatly admired, decries "Improvement" as the vulgar "idol of the age" with "many a victim" (230). Ventriloquizing her taciturn heroine, Austen puts these words (Cowper's and her own) in her heroine's mouth: "Cut down an avenue! What a pity! Does it not make you think of Cowper?" (83).

True to form, however, Austen wryly mocks the dominant trend in British landscape design. When it comes to the natural world, there was no "improving" what, in her opinion, was already perfect. In *Sense and Sensibility*, Marianne Dashwood operates as another proxy, even if the flawed Marianne needs to learn the value of self-discipline. Marianne is perplexed by the fact that Edward Ferrars prefers geometrically designed gardens (in the European style) *over* the rough and rugged picturesque style. No "nettles, or thistles, or heath blossoms" for him but orderly vistas and perspective (127). Marianne's excitement for wild nature bespeaks her unruly sensibility, or what Austen terms Marianne's "wildest anxiety" (200). Early in *Sense and Sensibility*, she confesses to her mother that hearing Edward recite poetry to her failed to kindle her excitement: "To hear those beautiful lines which have frequently almost driven me wild, pronounced with such impenetrable calmness, such dreadful indifference!" (55). From Eve K. Sedgwick's inflammatory "Jane Austen and the Masturbating Girl" to *Jane Austen and*

The Body, Marianne's alignment with highly eroticized forms of wildness—or what Sedgwick calls the character's "degeneracy" (109) and what Wiltshire calls her "love of the wild and romantic" (31)—has been exhaustively discussed. What has not been addressed is that Marianne sees Edward's aversion to dirt, which establishes him as an urban gentleman, a cleanly Londoner and *not* a dirty rube (hence, his complaint: "these bottoms must be dirty in winter"), and this is yet another reason why the pair would never make a happy and lasting couple (119). Marianne reacts with nothing but exasperation to Edward's inability to put dirt out of his mind: "How can you think of dirt," she sighs, "with such objects before you?" (119). This is a failed attempt on Marianne's part to make Edward feel the wild emotion so central to her sensibility.

The laconic Fanny, who has to work *less* hard than Marianne when it comes to checking her wildest of imaginations, embodies a similarly heightened sensitivity to her environs. Later in the narrative, she connects the sublimity of wild nature with her own moral sense since the sheer vastness of the land and sky helps to put her personal problems in perspective (135). Fanny observes the great Romantic dictum promulgated by William Wordsworth in *Lyrical Ballads* of 1789: "Let Nature be your teacher" (136). Playing the part of the humble pupil, she looks on wild nature in the same way that Elizabeth Bennet looks on, with wild amazement, at the park at Pemberley. The major difference between these scenes of nature-observation is that, whereas Elizabeth takes it all in nonverbally, Fanny wants her first cousin (and future husband) Edmund Bertram to imagine the wild and, in doing so, feel transported. Standing side-by-side in an open window, the two cousins look out over an "unclouded night" and the "deep shade of the woods" (135). Fanny "spoke her feelings," writes Austen (135). She proceeds to rhapsodize on what she calls the "sublimity of Nature," sublime enough that "all painting and music" and "poetry" can be left behind (135). At the emotional core of the scene's allusiveness is Act III, Scene V of *Romeo and Juliet* in which the eponymous couple talk excitedly about nightingales, clouds, and mountaintops. In Austen's version, the female speaker locates the source of her sensualized pleasure in non-human nature rather than in the male counterpart who stands excitedly beside her.

The wilderness in view from Sotherton Court helps to soothe Fanny even if she and her cousin triangulate their ungratified desires through the wildness enclosed by the casement. Mrs. Norris, Fanny's scold of an aunt, is also on the scene to monitor the cousins' behaviour. Despite these restraints, Fanny's rapturous excitement in seeing Sotherton (the seat of the Rushworth family) is identical to the pleasure Elizabeth feels when beholding Pemberley. Whether Fanny is experiencing a state of sexual arousal, which she displaces onto the *tabula rasa* of the natural world, or simply reversing the roles and playing the pedagogue to her know-it-all cousin, is hard to know. Regardless, an open window wiles her from polite conversation. She turns towards the night sky and dark woods to "tranquillize" emotions that are physically felt and transmitted between cousins: "his eyes soon turned like her's toward the scene" whereupon "Fanny spoke her feelings" (135). Though Austen stops short of using the word "wild," Fanny is *almost* wild insofar as she gets excited when imagining what lies in the dark outside the house: before she is told to leave her spot by the window, she feels "pleasure," then "glee," then "rapture" (135). Knowing that Edmund is more of a moralist than a sensualist, she directs this "glee" towards the educative, and remarks that more people would be transported "out of themselves" if they pondered "such a scene" (135). This is verbal catnip for Edmund who responds that those lacking Fanny's feelings for nature (clouds, woods, and the wildland) are to be pitied. The scene ends abruptly with such a person, Mrs. Norris, telling them to leave the window so that they don't catch cold. She serves another purpose, of course, which is to diffuse the wildness felt by Fanny and shared with Edmund.

As the owner of Mansfield Park, Fanny's future father-in-law is even more repressive than Mrs. Norris and even further removed from the wildness associated with Fanny's feelings. A slave-owning baronet and imperious patriarch, he has the last word when it comes to wildness, for he is the warden of the novel's titular prison and his niece is his inmate. He is also the enemy of feminine wildness, which he verbalizes in his denouncement of Fanny's character. By rejecting the advances of Henry Crawford, she invites her uncle's condemnation. In a long speech, he scorns Fanny as wild and unmannerly: "You think only of yourself . . . and are, in a wild fit of folly, throwing away from you such an opportunity of being settled in life, eligibly, honourably, nobly settled, as will, probably, never occur to you again" (324). In a long, patronizing diatribe that brings his niece to tears, he warns against the impropriety of wildness and ingratitude. A believer in female inferiority, Bertram is outraged that the niece whom he attempted to "improve" might marry based on genuine love and affection versus "being settled" (324). In other words, Fanny's bad decision-making (condemned as a sign of her wildness) has cost her the propriety—the "duty and respect"—that Bertram prizes above real love (324). Narratively, the ironic outcome in *Mansfield Park* is that the boy/man whom Fanny has loved since girlhood is Bertram's own son though the incest taboo matters less to him than Fanny's rejection of Crawford and his aristocratic status.

By the end of his monologue, Fanny has internalized Sir Bertram's opinion—she is more vulnerable to outside influences than Elizabeth Bennet in this respect—and Austen's use of exclamation points (used, earlier *and* deliberately, in her declarations of nature-love) reinforces the degree to which patriarchal power has come to override her prior attachment to the natural world: "Her heart was almost broke by such a picture of what she appeared to him; by such accusations, so heavy, so multiplied, so rising in dreadful gradation!" (325). Again, the rawness of feminine feeling is mediated by how an interpolated Fanny imagines that she is perceived by the novel's patriarch. Of particular interest to me is how the discourse of improvement in *Mansfield Park* joins one of the major motifs in Austen, that is, the project of female self-improvement. Kathryn Sutherland has gone so far as to designate Fanny Price the "object of Sir Thomas's experiment in improvement" (xxi), whereas the chief improver, according to Devoney Looser, is Austen herself who, with her own novels, supplants the earlier works of philosophy and conduct literature with which her female readers were familiar and, in doing so, improves their lives (45). Due to its elusive irony, her fiction lies somewhere between competing categories: romanticism versus realism, conduct literature versus Wollstonecraftian feminism, and, in this horticultural context, the wildness of the picturesque versus the humanized park (new and "improved"). Each of her bonneted heroines can be seen as a kind of "before" and "after," akin to the leather-bound "Red Books" that Humphrey Repton created for his clients; these were the original watercolours of the estate with his modification overlaying the original grounds. Sir Bertram is Repton's parallel in this way.

That said, irony is always a destabilizing tool of the literary trade, which makes it difficult to know where precisely her heroines stand in relation to Lagerspetz's equation: "*Cleanliness = Self-discipline = Progress*." Three of her beloved characters, Elizabeth, Catherine, and Fanny, are wild, dirty, and proudly so. Moreover, all three come to claim propriety for/as rural women without sacrificing some degree of their "wild manner." This is partly why, narratively, Elizabeth triumphs as the lady of Pemberley, the wife of Mr. Darcy, the queen of Derbyshire, whereas her immoderate and lustful younger sister, Lydia, is condemned to wander. Lydia's life with Wickham was "unsettled in the extreme," Austen writes in the conclusion to *Pride and Prejudice*, for the couple were "always moving from place to place in quest of a cheap situation" (384). Her older sister, by contrast, settles into Pemberley and continues the project of female improvement with her sister-in-law Georgiana as her pupil. The Austenian heroine needs the comfort that comes with being located in one place, which means that the feltness of place is not what Sir Bertram means by a woman feeling "nobly settled."

Austen not only sought to show how Romantic-era women were governed by the dichotomization of dirtiness and cleanliness, wild and self-controlled, but how wildness dislocates itself from the natural world when it comes to signify something more affective (i.e. fits and follies) and cognitive (i.e. "imaginations") in the lives of her heroines. To approach Jane Austen in this way is to recognize her novels, traditionally separated from the nature-writing of her contemporaries, as outgrowths of such discourses as landscape design and conduct literature. In the words of *Wilding* author Isabella Tree, "it is perhaps unsurprising that the Latin word for soil—'humus'—gives us 'human' and 'humility.' The soil is, quite literally, what grounds us" (290). Dirt grounds men and women in obvious ways, while cultural forces shape and misshape our original natures in less obvious ways. Austen understood this. Thus, Elizabeth, Catherine, and Fanny struggle to find the right balance between nature, or feeling wild, and culture, staying in control. Understanding

Austen in this way requires that we adopt a different footing. Her fiction satirizes the dominant notion that the true English lady is immaculately clean, spotless, and never wild. Far from being expunged, wildness is retained and managed in the same way that the English landscape was enclosed and improved upon by external forces. This is why dirt is no insignificant thing in Austen's fiction, for it serves to externalize a lady's innermost wildness, and, to the extent that dirt can never be completely removed from the social spaces reserved for English ladies and gentlemen, dirt is Austen's ironic reminder that inside every young lady lies a wild-woman.

Author's Note: I would like to thank The Jane Austen Society of North America, and its International Visitor Program, for their generous support of my research at the Jane Austen House and Museum and the Chawton House Library during the summer of 2019.

Notes

1 Singer contends that Austen's characters possess a peculiar resistance to the rawness of emotions, manifested through physical reactions, such as facial tics or shoulder-movement. Instead, their affective responses to others are modulated. This is further proof of what Singer identifies as the "mediated representations of emotions that are not so much fantasies as iterative constructions of affective worlds" (96). See Faflak's discussion of *Persuasion* and how the Musgrove girls' wildness is curbed (114–115).
2 For connections between wildness and emotional excess in Mary Shelley's fiction, see my *The Radical Ecology of the Shelleys* (2019): 154–190.
3 For Austen's life in Bath, known for its dampness, dirt, and mists, see Worsley (2017): 153–164.
4 In Austen's time, ladies wore what were known as "pattens," light-weight shoes used to elevate their feet above a dirty walkway.
5 For more on the influence of Repton and his predecessor, Lancelot "Capability" Brown, see Uglow (162–167).

Works cited

Austen, Jane. *Emma*. Ed. Kristin Flieger Samuelian. Broadview P, 2004.
———. *Jane Austen's Letters*. Ed. Deirdre Le Faye. Oxford UP, 1995.
———. *Mansfield Park*. Ed. June Sturrock. Broadview P, 2001.
———. *Northanger Abbey*. Ed. Claire Grogan. Broadview P, 2002.
———. *Persuasion*. Ed. Linda Bree. Broadview P, 1998.
———. *Pride and Prejudice*. Ed. Robert Irvine. Broadview P, 2002.
———. *Sanditon. Jane Austen's Manuscript Works*, edited by Linda Bree, Peter Sabor, and Janet Todd. Broadview P, 2013, pp. 319–385.
———. *Sense and Sensibility*. Ed. Kathleen James-Cavan. Broadview P, 2001.
Austen-Leigh, James Edward. *A Memoir of Jane Austen and Other Family Recollections*. Oxford UP, 2002.
Bate, Jonathan. "Culture and Environment: From Austen to Hardy." *New Literary History*, vol. 30, no. 3, 1999, pp. 541–560.

Carman, Colin. *The Radical Ecology of the Shelleys: Eros and Environment*. Routledge, 2019.

Carson, Rachel. *Silent Spring*. 1962. Houghton Mifflin, 1994.

Cowper, William. *The Works of Cowper*. The MacMillan Company, 1924.

Cronon, William. "The Trouble with Wilderness; or, Getting Back to the Wrong Nature." *Uncommon Ground: Rethinking the Human Place in Nature*. Norton, 1995, pp. 69–90.

Faflak, Joel. "Jane Austen and the Persuasion of Happiness." *Romanticism and the Emotions*, edited by Joel Faflak and Richard Sha. Cambridge UP, 2014, pp. 98–123.

Gentile, Kathy Justice. "Dirty Girls, Dirty Books, and the Breakdown of Boundaries in Jane Austen's Fiction." *Persuasions: The Jane Austen Journal*, vol. 39, 2017, pp. 57–69.

Heringman, Noah. *Romantic Rocks, Aesthetic Geology*. Cornell UP, 2004.

The Jane Austen Book Club. Directed by Robin Swicord, performances by Kathy Baker, Emily Blunt, and Maria Bellow. Hugh Dancy, 2007.

Johnson, Claudia L. *Jane Austen: Women, Politics and the Novel*. U of Chicago P, 1988.

Keats, John. *Keats's Poetry and Prose*. Ed. Jeffrey N. Cox. Norton 2009.

Kelly, Helena. *Jane Austen: The Secret Radical*. Knopf, 2017.

Lagerspetz, Olli. *A Philosophy of Dirt*. Reaktion Books, 2018.

Lau, Beth. "Teaching Jane Austen and the Male Romantic Poets." *Romantic Circles*. April 2015, https://romantic-circles.org/pedagogies/commons/austen/pedagogies.commons.2015.lau.html

Looser, Devoney. "(Re)Making History and Philosophy: Austen's *Northanger Abbey*." *European Romantic Review*, vol. 4, no. 1, 1993, pp. 34–56.

Mellor, Anne K. *Romanticism & Gender*. Routledge, 1993.

Richardson, Samuel. *Pamela*. 1740. Oxford UP, 2001.

Schülting, Sabine. *Dirt in Victorian Literature and Culture: Writing Materiality*. Routledge, 2016.

Sedgwick, Eve. "Jane Austen and the Masturbating Girl." *Tendencies*. Duke UP, 1993, pp. 109–129.

Seeber, Barbara. *Jane Austen and Animals*. Ashgate, 2003.

Singer, Kate. "Austen Agitated: Feeling Emotions in Mixed Media." *Jane Austen and Sciences of the Mind*, edited by Beth Lau. Routledge, 2018, pp. 95–114.

Tomalin, Claire. *Jane Austen: A Life*. Vintage Books, 1997.

Tree, Isabella. *Wilding: Returning Nature to Our Farm*. New York Review Books, 2019.

Uglow, Jenny. *A Little History of British Gardening*. Pimlico, 2005.

Wilshire, John. *Jane Austen and the Body*. Cambridge UP, 1992.

Wilson, Kim. *In the Garden with Jane Austen*. Frances Lincoln Limited, 2009.

Wordsworth, William. *Lyrical Ballads*, edited by M. Gamer and D. Porter, Broadview, 2008.

Worsley, Lucy. *Jane Austen at Home*. St. Martin's P, 2017.

10 "Wild above rule or art"

Volcanic luxuriance, subterranean terror, and the nature of gender in Ann Radcliffe's *A Sicilian Romance*

James Lesslie

Wild natures recur in the writings of Ann Radcliffe, emerging in descriptions of her protagonists' mental and physical states as well as representations of their environment, particularly the copious natural landscapes which many contemporary and subsequent commentators note as a remarkable feature of her work. Wild elements link, literally and figuratively, the human and organic realms in Radcliffe's fictions and her exploration of their potentially transgressive tendencies scrutinizes constructions of natural difference which underpinned social and political hierarchies in eighteenth-century Britain. Her novels dramatize debates about the status of women that predated the outbreak of the French Revolution in 1789 but intensified following the collapse of the *ancien régime* (Ellis, *History* 50). In these disputes the grounds of nature were invoked to justify a range of opinions, from radical principles of natural equality to hierarchies of naturally determined gender difference. In Radcliffe's novels these various perspectives are contested through characters and plots driven by conflicting impulses to maintain and exceed social, physical, and geographical boundaries. The tendency to excess that characterizes Radcliffe's writing has been extrapolated as an archetype of the gothic novel by modern critics, for whom the genre is characterized by an "excess of everything" (Kahane 46; Gamer 9). This appraisal turns a thematic concern of this anachronistically named genre into one of its defining tropes, an extrapolation of content that reflects critical practice during the 1790s and 1800s when the novels were made the subject of the very debates that they staged.

At this extratextual level the theme of excess persisted, as the proliferation of gothic fictions prompted fears about their potentially pernicious effects upon readers, a "moral panic" based on "a fantasy of corruption" which "hinged on the idea that women formed the principle audience for novels and the most custom for libraries" (Clery, *Rise* 97–98). This sense of profusion, or what eighteenth-century writings often termed "luxuriance," articulates the type of wildness that is the focus of this essay. It is a wildness that connotes natural abundance but also the potential to grow "rank," a point of excess which connects to another contemporary use of the word "wild" to describe uncultivated, and therefore unproductive, spaces ("luxuriant," adj.). The etymological roots of "luxuriance" link to the concept of luxury, a term applied to the burgeoning print world of the second half of the eighteenth century and the concerns generated by this profuse media, which were manifested in the novel's associations with "excessive consumption" and "unruly popular culture" (Clery, *Rise* 7; Ellis *History* 111). The gothic novel produced anxiety and censure among proponents of the very patriarchal values that it scrutinized. In turn, attempts to proscribe and dismiss these writings consolidated their perceived association with growing numbers of new producers and consumers, including women, which further troubled contemporary critics and perpetuated the genre's marginalization within canonical accounts of Romanticism until the twentieth century.

This essay focusses on the theme of wildness in Radcliffe's second novel *A Sicilian Romance* (1790) in order to consider how Radcliffe uses representations of wild natural and human impulses to critique contemporary constructions of gender difference. This analysis also shows how this often-overlooked early novel demonstrates the intertextual complexity of Radcliffe's compositions and her critical engagement with contemporary culture. Sicily was represented by eighteenth-century British writers as a site of natural abundance. Montesquieu's influential division of Europe into "northern" and "southern" nations on the basis of climatic and geographic characteristics in *The Spirit of Laws* (1750) had distinguished the south as a place where "nature has given much, and demands but little" (2: 23). While northern peoples were kept from degenerating into "barbarians" by the labour it took to make their terrain productive, nature's generosity in the south encouraged human "laziness," as little "industry" was required to support existence (2: 23). Montesquieu's theories about the effects of climate and geography upon national character influenced British conjectural historians and travel writers throughout the second half of the eighteenth century when southern Europe, particularly Italy, continued to be associated with profuse natural forces. Patrick Brydone's *A Tour through Sicily and Malta* (1773) generated renewed interest in Sicily and its landscape among British readers. In his opening letter to William Beckford, Brydone noted that Sicily was off "the old beaten track" of the typical grand tour and that "probably there were a variety of objects, not less interesting, that still lay buried in that celebrated island" (1: 2). The notion that Sicily might reveal "buried" materials presents the island as the site of hidden secrets, making it an apposite location for Radcliffe's mysterious narrative.

In *A Sicilian Romance* Radcliffe dramatizes the sense of conflict inherent in Brydone's representations of the Sicilian landscape, particularly the section on his journey to Mount Etna. Here, Brydone describes the volcanic terrain as a place where oppositional elements exist proximate to one another. Etna is said to "reunite every beauty and every horror" and "all the most opposite and dissimilar objects in nature," as Brydone observes "a gulph, that formerly threw out torrents of fire now covered with the most luxuriant vegetation" (1: 186). The landscape's wildness is given a dual character as the source of both abundant pleasures and danger:

> Here the ground is covered with every flower; and we wander over these beauties, and contemplate this wilderness of sweets, without considering that hell, with all its terrors, is immediately under our feet; and that but a few yards separate us from lakes of liquid fire and brimstone. (Brydone 1: 186)

Brydone's admiration of the profuse "beauties" that cover Etna's surface also registers a sense of anxiety about its wild condition. These "sweets" distract the mind from the "hell" below that both enables and endangers them. This precariously contained elemental power might at any time exceed its limits and erupt, yet despite its destructive potential it is also generative. Brydone's description of "this wilderness of sweets" references *Paradise Lost* (1667) and the view of Eden seen by the angel Raphael as he arrives "Into the blissful field" in Book V and observes:

> A wilderness of sweets; for nature here
> Wantoned as in her prime, and played at will
> Her virgin fancies, pouring forth more sweet,
> Wild above rule or art; enormous bliss. (V.294–297)

The "hell" beneath Brydone's feet is also drawn from "the lake with liquid fire" that Satan discovers in Milton's hell, which itself resembles "the shattered side / Of thund'ring Etna" (I.229, 232–233). By incorporating both Milton's Eden and his hell in this portrayal of Etna, Brydone suggests that the island's paradisiacal "beauties" are precariously united with its subterranean terrors and presents the volcano as a productive yet potentially ruinous force.

The "wanton" and "overgrown" wildness of Milton's garden differentiates it from previous literary depictions of Eden (Evans 248). Nature is beyond "rule" and Adam and Eve's inability to control its luxuriance problematizes prelapsarian man's place at the apex of an ordered creation as God's representatives on earth. Instead, nature's transgressive disorder lies latent in paradise, anticipating and prefiguring the fall of Adam and Eve, who are literally and figuratively intertwined with the flora that they both tend and cultivate. Their transgression, particularly Eve's, is anticipated by the wild condition of the natural space that they inhabit. Milton constructs his luxuriant Eden and its unruly female inhabitant via mythical allusions which identify them with established literary geographies of natural profusion. These include Enna in Sicily where, Brydone recalls, Proserpine was allegedly "carried off" by Pluto to the underworld (2: 245). Brydone notes that Enna, like Etna, is "covered with a variety of flowers at all seasons of the year," before concluding his discussion of Proserpine's abduction by quoting Milton's description of Eden as surpassing even "that fair field": "Of Enna, where Proserpine gathering flowers, / Herself a fairer flower, by gloomy Dis / Was gathered" (2: 245–246). Mandy Green argues persuasively that the flowers which attract and embody the doomed Proserpine identify her with Eve in Book IX, when Satan spies her "stooping to support / Each flow'r of slender stalk," while she is herself described as "fairest unsupported flower" (IX.427–428, 432; Green 920–921). Seen through the diabolic male gaze, Proserpine and Eve present ambivalent images of female agency interconnected with the naturally profuse landscapes that they inhabit. For Green the "gathered" Proserpine represents female passivity in the face of male aggression, while Eve is complicit in her own downfall and becomes "both victim and agent of the tragic process" (920). The wildness of Eden and its luxuriant flora can therefore be read as figurative analogues of Eve's problematic independence and the complex "subject position" that Milton gives her in *Paradise Lost* (Green 922).

While some eighteenth-century writers understood Milton's Eve as a naturalized emblem of female passivity, in *A Sicilian Romance* Radcliffe activates Eve's ambiguous analogical relationship with wild nature as both a liberated and subjugated figure. The novel explores the Miltonic conflict between the naturally independent, inquisitive female subject and external perspectives that posit a more passive female nature. These contrasting views of female identity were subject to debate during the second half of the eighteenth century as the notion of naturalized gender difference was contested by radical writers. In *A Father's Legacy to his Daughters* (1761) the Scottish physician John Gregory selected the seemingly demure figure of Eve to exemplify "the most perfect simplicity of heart and manners" (46). In *Sermons to Young Women* (1775) James Fordyce described the poem's "picture of Eve, in her state of innocence . . . as the model of woman most amiably feminine" (1: 97). Fordyce obscured the transgressive connotations of Eve's figurative association with the wild landscape of Eden to source an image of decorous female passivity._Mary Wollstonecraft critiqued Fordyce's sermons in *A Vindication of the Rights of Woman* (1792) for attempting to cajole women "into virtue by artful flattery and sexual compliments" (210). Wollstonecraft also criticized Milton's representation of Eve, "our first frail mother," as a model of "docile blind obedience" that encourages women "only to render ourselves as gentle domestic brutes" (33). In *A Sicilian Romance*, the contrasting character of the Sicilian landscape that Brydone observes around Etna provides a metaphorically apposite geography for the contested determination of female identity between containment and excess that the novel depicts. Radcliffe questions the extent to which women's natural powers can be confined and what the consequences of such confinement might be.

The outbreak of the French Revolution in 1789 gave the debates about individual liberties dramatized in *A Sicilian Romance* an urgent political and ideological context. While Radcliffe's novel does not directly engage with contemporary events, its Sicilian setting permits the thematic tensions present at geographical and personal levels to be considered within a broader political perspective. Radcliffe's predecessors made explicit connections between Sicily's volatile landscape and its political situation. Brydone's images of repressed natural forces on the brink of eruption contain an insurrectionary aspect that he extended to his description of its populace. He used the region's natural abundance as a damning indictment of its rulers, wondering in disbelief that "any government should be able to render poor and wretched, a country which produces, almost spontaneously, every thing that even luxury can desire" (Brydone 2: 57). For much of the eighteenth century Sicily was under Spanish dominion, and Brydone's condemnation of the island's governance reveals an anti-Hispanism that extends these sufferings to anywhere under "the Spanish yoke," both here and "on t'other side of the globe" (2: 27). Despite this oppression, Brydone finds the Sicilians to be "people of great sensibility, with high notions of honour and liberty," who "would readily embrace any plausible scheme, to shake off their yoke" (2: 285). He presents Sicily as the site where naturally profuse and potentially disruptive elements are suppressed by powers that may not be able to contain them indefinitely. The wild fecundity of the island's landscape thereby becomes analogous to its inhabitants' restive character, which is in turn linked to their "great sensibility." Brydone's alignment of heightened "sensibility" with insurrection indicates the transgressive potential of raised emotional receptivity and its ability to meld personal and political registers. A decade later, Henry Swinburne justified his account of a region that had "been so often described": "In the southern parts of Italy, where the elements ferment with more than ordinary violence, where changes in government have succeeded each other with uncommon rapidity, the variations are more precipitate, the effects more striking" (1: iv). These travel narratives presented Sicily as a place where human control over natural, personal, and political order was tentative and these "elements" were frequently on the verge of exceeding their boundaries, expressing "the hyperbolic potential of immoderation in the warm South" (Chard 5). Radcliffe's novel presents the island as an imagined geography where wild nature is both a place of danger and the site of potential liberation as the boundaries imposed by oppressive power structures are transgressed.

The tendency towards excess that Brydone and Swinburne located in the Sicilian landscape and its people is evident in Radcliffe's introduction of her novel's heroine, Julia, who is said to be possessed of an "extreme sensibility," a "warm" temper, and an "ardent" imagination (1: 7). While these traits reflect stereotypes about southern Europe, the description of Julia's "extreme sensibility" also references sentimental novels like Rousseau's *Julie, ou La Nouvelle Héloïse* (1761), which shared similar preoccupations about excessive passions and the limits of socially sanctioned behaviour. Sentimental and travel writings of the late eighteenth century privileged "emotional responsiveness," a quality that Julia displays in abundance, while also marking such susceptibility as a "feminized attribute" (Chard 36). On one level, the increased cultural prominence of these "feminized" figures gave female writers a means to articulate progressive opinions. Authors could use literature to oppose "social ills" with the redemptive possibilities of "social affection" and common humanity (Todd 60). They could also, as Radcliffe does in *A Sicilian Romance*, present female characters whose heightened "sensibility" leads to self-education, self-expression, and the exploration of their environment, showing that these endeavours are as naturally undertaken by women as they are by men. Nonetheless, the threat of excess continued to be a point of tension within both sentimental novels and the contemporary response to them. Claudia L. Johnson shows criticism of sentimental writing rising after 1789, when the sympathetic inclinations described in the literature of sensibility became associated with revolutionary radicalism and were accused of destabilizing hierarchical values and legitimizing personal, and consequently public, disorder (6). In turn, Edmund Burke and other male writers deployed sentimental tropes to maintain gendered boundaries, suggesting that "only men have legitimate access to the discourse of the heart" and further marginalizing "the women whose distress occasions their affective displays from enjoying any comparable moral authority by representing their affectivity as inferior, unconscious, unruly, or even criminal" (Johnson 14). By the 1790s the anxieties generated by the articulation of powerful emotions, especially in women, created an "impasse of sensibility," wherein novels frequently detail "feelings that are declared overwhelming, but that are consistently brought under command and suppressed" (Ellis, *Politics* 217). Gothic novels developed the dynamic tension between expression and repression articulated in the literature of sensibility, locating this struggle within more volatile historical and geographical scenarios. Against such unstable backgrounds these conflicting impulses were heightened and could take on a violent, disturbing character.

In *A Sicilian Romance* Julia's "extreme sensibility" propels a journey from her childhood home through wild and subterranean parts of Sicily where she attempts to elude the continual threat of confinement. She and her sister initially inhabit their father's castle, supervised by Madame de Menon, whose "particular care" it is "to counteract those traits in the disposition of her young pupils, which appeared inimical to their future happiness" (1: 8). Julia is practised in the aesthetic contemplation of nature from a young age, indulging "an early taste for books" in "a small closet" whose windows look "upon the sea . . . the dark rocky coast of Calabria" and "a prospect of the neighbouring woods" (1: 10–11). However, the pleasures of this "prospect" are complicated by the conditions that enable it. The "eminence" which allows Julia's eyes "an almost boundless range of sea and land," including "the wild and picturesque scenery of Sicily" and the "grand and sublime" Mount Etna, is circumscribed (1: 12). It is revealed that Maria de Vellorno, her father's second wife, has "employed all her influence over the marquis to detain" Julia and her sister Emilia "in retirement" and that "they had never passed the boundaries of their father's domains" (1: 13). The repressed natural and human energy that Brydone and Swinburne noted in their accounts of Sicily is evident in Radcliffe's representation of Julia's "ardent" imagination and its confinement within externally imposed "boundaries." Her subjugation is emphasized by the dilemma she faces as the novel progresses. She falls in love with the Count de Vereza, also known as Hippolitus, but her father commands her to marry the Duke de Luovo in a match that he believes will bring him "wealth, honour, and distinction" (1: 127). The conflict dramatized here arises from a tension between an older generation that espouses hereditary and patriarchal power and a younger generation that uses the language of sensibility and natural rights. Hippolitus urges Julia to "Fly . . . from the authority of a father who abuses his power, and assert the liberty of choice, which nature assigned you" (1: 140). The concepts that Hippolitus introduces here—the abuse of hereditary power, freedom of choice as "assigned" by "nature"—were central to contemporary debates about the incipient revolution in France. In *Reflections on the Revolution in France*, published the same year as *A Sicilian Romance,* Burke reviles the progress of revolution as a dangerous assault upon hereditary principles which he sees as essential to the persistence of successful governments and societies. Julia defies her father's "authority" and flees her home, abandoning a marriage based on "the oppressive usages of custom," which in eighteenth-century British common law largely passed women and their property into the legal ownership of their husbands (Clery, *Rise* 126). Her initial perspective on the island's scenery expresses a power structure which curtails her natural freedoms and denies her sensibility. In order to pursue her heart's inclinations, it is necessary for her to move beyond its boundaries.

Julia's escape from her father's property shifts the action of the novel into the Sicilian countryside. Radcliffe draws on the concept of the richly productive south and immerses her female characters in variegated and dramatic natural scenery experienced at first hand. Following Julia's escape, Madame de Menon quits the castle, travelling on her way to retirement in a convent. Arriving at a small village for the night, she is tempted to walk among "the grotesque beauty of the surrounding scenery," and follows "a stream" which is "lost at some distance amongst luxuriant groves of chesnut [*sic*]" before entering "the shades" (2: 1). In the novel's profuse Sicilian setting these images of natural luxuriance are "congenial" to this wandering woman (2: 1). The trees create a "pensive gloom," while Madame de Menon's thoughts are pleasingly "affected by the surrounding objects" and she is "insensibly led on" in a manner that suggests an unconscious correspondence between herself and her environment (2: 1–2). This phrase recalls Robert Lowth's *Lectures on the Sacred Poetry of the Hebrews* (1753) which describes "a kind of continued negation" that is a feature of these ancient writings: "the boundaries are gradually extended on every side, and at length totally removed; the mind is insensibly led on towards infinity, and is struck with inexpressible admiration" (Lowth

2: 354). Fiona Stafford argues that Lowth's work influenced the style of James Macpherson's Ossian poems, which is in turn identifiable in the anthropomorphic conflation of scenery and emotion in Radcliffe's novels (89). Macpherson's verses, alongside the apparently primitive Highland culture that they both represented and appropriated, disturbed eighteenth-century hierarchies of cultural authority on account of their emergence from uncultivated literary geographies and their unverifiable textual authenticity, provoking suspicions of forgery which were also associated with gothic romances following Horace's Walpole's "hoax" in the 1764 edition of *The Castle of Otranto* (Gamer 60). The attempt to marginalize and diminish these dubious utterances is evident in Samuel Johnson's repeated attacks upon Ossian. When asked by Hugh Blair in 1763 "whether he thought any man of a modern age could have written such poems," he replied: "Yes, Sir, many men, many women, and many children" (Boswell 396). Johnson derides these compositions by associating them with uncultivated and improper literary practitioners: namely women and children. As Michael Gamer notes, this "Johnsonian" prejudice persists in reviews of the gothic novel during the 1790s, and it is perpetuated in modern descriptions of Radcliffe as a "one of the most 'literary,' one might even say secondhand [*sic*], of important novelists" (34; Dekker 92).

Radcliffe's representation of Madame de Menon's interaction with the Sicilian landscape adapts her various sources to depict a woman's creative interaction with wild nature, revealing the allusive construction of her fictions as a critical process rather than mere "literary plundering" (Clery, *Women's Gothic* 56). Radcliffe enacts Lowth's model of the sublime in the profuse Sicilian landscape of Brydone's travel narrative and, as we will see, adjusts masculine tropes found in eighteenth-century loco-descriptive poetry. She presents the experience of wild nature as a reflexive process, with Madame de Menon's physical progression through space portrayed via the sequential revelation of visual images. Nature is not framed but instead draws the spectator along and is revealed heterogeneously as it impacts on her consciousness, which projects its affective responses back upon the scenery. The sight of "wild and grotesque rocks" leads her to contemplate nature in "her vast magnificence," which in turn raises "the beholder to enthusiasm" (2: 2). This gendering of "Nature" as female at the point of Madame's mental elevation implies a correspondence between these two figures. In her analysis of eighteenth-century picturesque theory Vivien Jones has observed that the female gendering of nature was intrinsic to the discourse's "voyeuristic" aesthetics, as it was frequently predicated on the male viewer's pursuit of nature's feminine beauties (128). Radcliffe's depiction of a female character's physical and mental immersion in this picturesque and sublime landscape modifies the androcentric power dynamic encoded in contemporary theories of aesthetics, presenting a reflexive process of mutual influence rather than the pursuit of mastery.

Immediately after this gendering of the natural world, another female figure, "Fancy," enters the scene and unites the landscape in anthropomorphic communion with its female "beholder":

> Fancy caught the thrilling sensation, and at her touch the towering steeps became shaded with unreal glooms; the caves more darkly frowned — the projecting cliffs assumed a more terrific aspect, and the wild overhanging shrubs waved to the gale in deeper murmurs. The scene inspired madame with reverential awe, and her thoughts involuntarily rose 'from Nature up to Nature's God.' (2: 2)

In *The Seasons* James Thomson, the author Radcliffe most references after Shakespeare, situates the masculine poet in a "mid-wood Shade" like the one Madame de Menon passes through before describing the process of poetic inspiration ("Summer," line 9). Here, "Fancy" enables the "raptur'd Glance / Shot on surrounding Heaven to steal one Look / Creative of the Poet" ("Summer," lines 16–19). While for Thomson's male narrator "Fancy" connects the "Mortal" to his surroundings in a moment of "Ecstasy," in *A Sicilian Romance* Madame de Menon's "Fancy" establishes a woman's creative connection with the scenery, embellishing the landscape with affective detail and anthropomorphic projection ("Summer," lines 16, 20). Caves "frowned," the "towering steeps" are "shaded with unreal glooms," the cliffs are rendered "more terrific," and the "shrubs" respond to the "gale" with human gestures as they wave and murmur. This identification between a woman and wild nature results in a heightened reality, a state where the scenery and the viewer's imagination engage in a dynamic creative process. The ultimate effect for Madame de Menon is a moment of divine reflection, in which the mind moves from the contemplation of "Nature up to Nature's God," a pattern derived from Pope's *An Essay on Man* (IV.327). Through this moment of epiphanic communion emerging from the interaction between Madame de Menon, a female "Nature," and the similarly gendered "Fancy," Radcliffe connects women, the imagination, and wild nature and posits a moment of sublime inspiration in what Thomson presented as a traditionally masculine domain. As Madame de Menon continues to be led by the anthropomorphically murmuring scenery, Radcliffe reveals a human "voice of liquid and melodious sweetness" rising "from among the rocks." The "sweet warbler" is eventually identified as a "peasant girl" picturesquely immersed in the scene—"seated on a small projection of the rock, overshadowed" by trees (2: 3). This "peasant girl" is Julia in disguise, Madame de Menon having been "insensibly led on" by nature, perhaps even "Nature's God," to discover her. Julia's exchange of the detached prospects of the castle for a more immersed relationship with nature has involved a superficial alteration of her social position, symbolically registered in her disguise as a "peasant." In temporarily abandoning her status, Julia discovers an affective kinship with natural scenery demonstrated by the way in which her lament for the apparent death of her lover Hippolitus is "repeated" by the "rocks" surrounding her "with an effect like that of enchantment" (2: 3). For Julia, like Madame de Menon, wild nature offers a space of self-discovery and expression. Nonetheless, the fact that Julia must disguise and sequester herself in these spaces affirms the continued danger involved in her attempt to escape the socially sanctioned limits of female existence and shows that women's attempts at autonomy rarely go uncontested.

While these sublime scenes offer a space for female individuation, elsewhere in the novel the unruly forces associated with Sicily's wild landscapes appear to threaten Radcliffe's heroine. The dangerous elements of the Sicilian landscape have a masculine analogue in the banditti that feature in the narrative. These men manifest the violent energies of the island and are associated with its subterranean recesses, continuing its mythic associations with baleful male figures like Dis and Satan who appeared on the surface to steal women to the underworld. As Julia continues her flight, Hippolitus searches for her. In a ruined monastery he discovers "a group of men, who from the savageness of their looks, and from their dress, appeared to be banditti" (2: 126). Radcliffe's depiction of the savagery of these figures maintains the link that Brydone portrayed between the wild elements of the Sicilian countryside and violent human impulses which share the dangerous alterity of the spaces they inhabit. Hippolitus moves further into the "secret recesses of the pile" and discovers the body of an unconscious woman, later revealed to be Julia, being held captive (2: 128). He hides and watches as two banditti fight to the death over their prize, with the victor about to rape her prior to Hippolitus's intervention. Julia's attempt to escape a union forced upon her by the precepts of androcentric society leads her to the point of violation by savage forces who occupy Sicily's wilder, marginal spaces. The seemingly opposed conditions of barbarity and cultivation merge in a masculine perspective that reveals the rapacious impulses underlying patriarchal convention.

Radcliffe's association of her female characters with these sublime and subterranean scenes situates them within the ambiguous framework of the Sicilian landscape's tendency to profusion and disorder, positing an image of female autonomy that is nonetheless shaped by eighteenth-century poetic and aesthetic conventions. Even as Radcliffe adjusts these tropes to posit a notion of female creativity and self-expression, the voyeurism implicit in contemporary identifications of women as picturesque objects is potentially reinscribed by the conflation of women with landscape. During the revolutionary period, British women's efforts to create a liberated space within androcentric discourses could provoke further attempts to contain such gestures of self-assertion, and their exposure of patriarchy's inherent iniquities prompted the reaffirmation of this system as being based on fundamental, natural principles. For example, in *The Unsex'd Females* (1798) Richard Polwhele condemns Wollstonecraft and other radical women writers for advocating "liberty's sublime views" (12). He tellingly portrays revolutionary politics as "wild," described in Edenic terms as a "garden tempting with forbidden fruit," and identifies women's articulation of progressive principles with Eve's original transgression (15). Polwhele presents his attack on the idea of gender equality as a defence of women's "modest Virtue," a trait in keeping with the "weakness" he designates as their proper condition according to the natural "distinction of the sexes" (39, 18, 25). In Polwhele's reactionary justification of hierarchical society, the masculine custodian of women's virtue is also the self-appointed authority maintaining the boundaries of proper female expression. In Radcliffe's novel the perspectives of protector and captor blur when Hippolitus discovers the banditti's female hostage. His vision of this unconscious figure presents an aestheticized, objectified view of the female body which renders the identification of women and nature ambiguous in its association of wild elements with both containment and transgression.

Radcliffe's representation of Julia at this climactic moment in her narrative mobilizes the conflicting energies embodied in the Sicilian landscape to show the persistence of repression within patriarchal power structures while also presenting a provocative association between women and wild nature that both confirms and resists these perspectives. The discovery of her unconscious form is seen from Hippolitus's viewpoint:

> The count had advanced several steps before he perceived an object, which fixed all his attention. This was the figure of a young woman lying on the floor apparently dead. Her face was concealed in her robe; and the long auburn tresses which fell in beautiful luxuriance over her bosom, served to veil a part of the glowing beauty which the disorder of her dress would have revealed. (2: 132)

Radcliffe's depiction of Julia as a passive "object" has notable parallels with Milton's portrayal of Eve at two key points in *Paradise Lost*. The introduction of Adam and Eve in Book IV pays particular attention to their hair, describing Adam's "hyacinthine locks" which "manly hung" and stating that Eve

> as a veil down to the slender waist
> Her unadorned golden tresses wore
> Disheveled, but in wanton ringlets waved
> As the vine curls her tendrils. (IV.301, 302, 304–307)

In both scenes "tresses" act as a "veil" covering the upper body, while the "wanton" and unruly nature of Eve's hair is shared by the "luxuriance" of Julia's. In *Paradise Lost*, Eve's "tresses" are figuratively intertwined with the profuse condition of the garden via comparison to the "vine" which, several lines earlier, is itself described as "luxuriant" (IV.260). Milton's detailing of Eve's hair conveys exuberant female growth, a characterization that complicates the poem's depiction of the different capabilities conferred by Adam and Eve's respective genders: "For contemplation he and valor formed, / For softness she and sweet attractive grace" (IV.297–298). Here Eve, like Radcliffe's Julia, occupies the position of an "object" whose alluring features are there for the "contemplation" of the male eye. Milton refers again to Eve's "tresses discomposed" at the start of Book V, after Satan has spent the night at her ear attempting to reach the "organs of her fancy" and she dreams of eating from the tree of knowledge (V.10; IV.802). Eve then relates her dream to Adam, who advises her that "fancy," in the absence of reason, produces "Wild work" (V.112). The dream's prefiguring of Eve's ultimate transgression indicates the threat of these wild forces: Adam's advice to Eve suggests that female "fancy" requires regulation by a male authority, an assertion which nonetheless compounds its transgressive associations. In *A Sicilian Romance* Julia's attempts to follow her natural impulses and exceed her physical and mental boundaries lead to her subterranean confinement, where her "glowing beauty" is rendered an object for the banditti's and Hippolitus's contemplation. Radcliffe presents an ambiguous figuration of her heroine which activates notions of female passivity alongside images of "luxuriance" that both encourage and elude the possessive eye of the masculine observer. Her conflicting identification of women with wild natural forces in *A Sicilian Romance* mimics the "paradoxical formulation of female sexuality" advocated by late eighteenth-century moralists, who prescribed female modesty as protector against passions which it simultaneously advertised and provoked (Poovey 15, 19). This image of Julia also displays a tension identified by Diane Hoeveler in her assessment of the "female gothic" as "both subverting and reifying postures of complaisancy and acquiescence on the part of women" (xvi).

The historical depth of patriarchal repression is highlighted by the denouement of *A Sicilian Romance* which reveals another female figure imprisoned, like Proserpine, in the island's depths. Having escaped the banditti, Julia and Hippolitus are set upon by her father's men and she is forced to retreat into the "inmost recesses" of a "cavern" (2: 154). As she progresses further through this subterranean landscape, she discovers a series of doors which lead to further "dark" passages before eventually emerging into a "small room" (2: 158). Here, she discovers her mother, long presumed dead but in fact imprisoned by Julia's father to facilitate his marriage to Maria de Vellorno. Radcliffe draws upon the mythical history of Sicily as well as its representation in eighteenth-century travel literature to present an archetypal tale of female entrapment that also contains the possibility of liberation. The parallels between the mother's underworld confinement and the myth of Proserpine show how masculine power is predicated on the forcible delimitation of female autonomy, while the figurative presence of Eve both acknowledges and critiques contemporary models of female passivity by suggesting the transgressive potential that they sought to proscribe. Radcliffe's representations of epiphanic moments of female self-discovery amidst the wild Sicilian countryside offer alternative images of independent, spontaneous growth and show aesthetic space as something that women can shape rather than merely inhabit, an act that the heroines of her later novels will perform with increasing sophistication.

The lapsarian motifs that Radcliffe incorporates in *A Sicilian Romance* have since been applied to her literary reputation which, after the "treatment" of her next novel *The Romance of the Forest* (1791) as "a perfect expression of its form," presents "a fall from this edenic position" (Gamer 71). While Radcliffe's "prestige remains . . . remarkably high until at least 1810," after the publication of *The Mysteries of Udolpho* (1794) the association of the wider genre of gothic romance with excess is applied, albeit more sparingly, to her works (Gamer 71). By the end of the eighteenth century the content of Radcliffe's novels is often blurred with that of her successors and imitators to present a monstrous mass of romances that provokes reviewers, including Samuel Taylor Coleridge, to increasingly vexed denunciations of the genre and its influence on women writers and readers. The "stigmatization" of gothic fiction from the 1790s has itself been posited in gendered terms, as part of a process in which "canonical romanticism . . . based upon masculine tropes" suppresses (while at the same appropriating) a genre that is pejoratively characterized as feminine (Gamer 31, 5). The work of feminist scholars since the 1970s has done much to rebalance this canonical narrative and studies of the female gothic have rightly returned women writers to their central place in literary culture at the end of eighteenth century, a milieu in reality not as stratified or patriarchal as masculinist critics may have wished. Early feminist accounts of what Ellen Moers dubbed the "Female Gothic" "involve the construction of an alternative, repressed female tradition" which made the "subordination of women in society . . . the constant theme of their literary work" (Moers 90; Clery, *Women's Gothic* 2). In recent decades scholars have questioned the "dualism of male and female traditions" whose "gender-complementary" readings of gothic texts risk reinscribing "rigid gender boundaries" which were in fact far from stable at the end of the eighteenth century (Clery, *Women's Gothic* 2; Craciun 22). In this context, the censorious response of eighteenth-century writers makes "the real danger presented by the Gothic's temporary anomalies . . . visible in the anxieties of its critics" (Craciun 132). Rather than revealing the effective suppression of these texts' transgressive ambiguities, these increasingly anxious writings show that the gothic continued to disrupt naturalized models of gender difference. The works of later writers like Charlotte Dacre have been shown to describe female passions and violence that far exceed Radcliffe's "embattled" and "reactive" heroines, presenting "women characters who systematically perform actions 'unnatural' for women (such as dominate, assert, desire, aggress, and kill), thereby destabilizing the categories 'feminine' and 'female'" (Craciun 153). Such distinctions suggest that Radcliffe's depictions of women's encounters with wildness were perhaps not transgressive enough, reminding us that in reactionary misogynist invectives like Polwhele's *The Unsex'd Females* (1798), Radcliffe was offered as an example of a writer whose work did not exceed the proper bounds ascribed to her sex. Nevertheless, if the moralistic conclusion to a novel like Dacre's *Zofloya; or The Moor* (1806) does not undermine the "temporary anomalies" that precede it, then Radcliffe's tendency to resolve her fictitious dramas "by tactics of conciliation or evasion" should not be allowed to settle the ambiguities that emerge in her interconnected depictions of unruly human and organic natures (Clery, *Rise* 137).

In *A Sicilian Romance* Radcliffe critiques the construction of the "ideal" woman as unnatural and dangerous on account of the restrictions it places on women's bodies and imaginations (Craciun 135). She posits a version of womanhood that is subtly transgressive and possessed, like Milton's Eve, with an innate impulse to outgrow the strictures placed upon it. The novel's conclusion arguably attempts to contain this disorderly energy, yet the proliferation of Radcliffe's female heroines in her subsequent novels as well as in the imaginations of her readers and imitators continued to disturb contemporary orthodoxies around gender and culture well into the nineteenth century. The rumour, on her death in 1826, that Radcliffe's life ended in madness due to the "excessive use of her imagination in representing extravagant and violent scenes" is telling in this instance (Clery, *Rise* 196–197). This tale enacts upon the author the confinement that her heroines seek to escape on account of their "extravagant" imaginations, revealing the persistence of the repressive impulses towards female creativity and liberty that her writings question. Furthermore, it acknowledges her novels' transgressions of normative cultural boundaries by designating their author as aberrant and sequestering her at the margins of society. Like Wordsworth's dismissal of what he called the "Radcliffe school" (232) or the identification of gothic novels with circulating libraries, this institutionalization of Radcliffe's work reveals anxieties about its subversive potential and isolates it from the heterogenous literary discourses with which it engages. Just as Dacre's radical challenge to conceptions of gender difference is more fully revealed when situated in relation to the works of Matthew Lewis and Marquis de Sade, so Radcliffe's writings need to be considered in terms of her interactions with predecessors and contemporaries regardless of their gender—from Milton and Brydone to Ossian and the speculative histories of the Scottish Enlightenment—as well as their discernible impact on successors like Lewis, Dacre, and Walter Scott. When Radcliffe's work is resituated within this literary "continuum" (Craciun 131), in which generic and chronological categories such as enlightenment, gothic, and romantic are understood as contingent rather than definitive, her participation in revolutionary-period debates about political and cultural authority predicated on conflicting understandings of the relationship between civilization and the natural world is made more fully apparent.

Works cited

Boswell, James. *Boswell's Life of Johnson*, edited by George Birkbeck Hill, revised by L. F. Powell, vol. 1. Clarendon P, 1934.

Brydone, Patrick. *A Tour through Sicily and Malta*. 2nd ed. London, 1774.

Chard, Chloe. *Pleasure and Guilt on the Grand Tour: Travel Writing and Imaginative Geography 1600–1830*. Manchester UP, 1999.

Clery, E. J. *The Rise of Supernatural Fiction, 1762–1800*. Cambridge Studies on Romanticism 12. Cambridge UP, 1995.

———. *Women's Gothic: From Clara Reeve to Mary Shelley*. Northcote House, 2000.

Craciun, Adriana. *Fatal Women of Romanticism*. Cambridge UP, 2003.

Dekker, George G. *The Fictions of Romantic Tourism: Radcliffe, Scott, and Mary Shelley*. Stanford UP, 2005.

Ellis, Markman. *The History of Gothic Fiction*. Edinburgh UP, 2000.

———. *The Politics of Sensibility: Race, Gender and Commerce in the Sentimental Novel*. Cambridge UP, 1996.

Evans. J. M. *Paradise Lost and the Genesis Tradition*. Clarendon P, 1968.

Fordyce, James. *Sermons to Young Women*. vol. 1, London, 1766.

Gamer, Michael. *Romanticism and the Gothic: Genre, Reception, and Canon Formation*. Cambridge UP, 2004.

Green, Mandy. "The Virgin in the Garden: Milton's Ovidian Eve." *Modern Language Review*, vol. 100, no. 4, October 2005, pp. 903–922.

Gregory, John. *A Father's Legacy to His Daughters*. London/Edinburgh, 1774.

Hoeveler, Diane. *Gothic Feminism: The Professionalization of Gender from Charlotte Smith to the Brontës*. Pennsylvania State UP, 1998.

Johnson, Claudia L. *Equivocal Beings: Politics, Gender, and Sentimentality in the 1790s: Wollstonecraft, Radcliffe, Burney, Austen*. U of Chicago P, 1995.

Jones, Vivien. "'The Coquetry of Nature': Politics and the Picturesque in Women's Fiction." *The Politics of the Picturesque*, edited by Stephen Copley and Peter Garside. Cambridge UP, 1994, pp. 120–144.

Kahane, Claire. "Gothic Mirrors and Feminine Identity." *The Centennial Review*, vol. 24, no. 1, Winter 1980, pp. 43–64.

Lowth, Robert. *Lectures on the Sacred Poetry of the Hebrews*. Translated by G. Gregory, vol. 2. London, 1787.

"luxuriant, adj." *OED Online*. Oxford UP, March 2020, www.oed.com/view/Entry/111506.

Milton, John. *Paradise Lost*, edited by Scott Elledge. Norton, 1993.

Moers, Ellen. *Literary Women*. W. H. Allen, 1977.

Montesquieu. *The Spirit of Laws*. Translated by Thomas Nugent, vol. 2. London, 1750.

Polwhele, Richard. *The Unsex'd Females*. New York, 1800.

Poovey, Mary. *The Proper Lady and the Woman Writer: Ideology as Style in the Works of Mary Wollstonecraft, Mary Shelley, and Jane Austen*. U of Chicago P, 1984.

Pope, Alexander. *Essay on Man*, vol. 4. London, 1734.

Radcliffe, Ann. *A Sicilian Romance*. London, 1790.

Stafford, Fiona J. *The Sublime Savage: A Study of James Macpherson and The Poems of Ossian*. Edinburgh UP, 1988.

Swinburne, Henry. *Travels in the Two Sicilies*, vol. 1. London, 1783.

Thomson, James. *The Seasons*. London, 1744.

Todd, Janet, *Sensibility: An Introduction*. Methuen, 1986.

Wollstonecraft, Mary. *A Vindication of the Rights of Woman: with Strictures on Political and Moral Subjects*. London, 1792.

11 "A strange unearthly climate"

James Hogg's tale of the Arctic wild

Robert W. Rix

In his "Fact Book," Henry David Thoreau noted the following: "Wild—past participle of to will—self-willed" (qtd. in Turner 37). This is not a genuine etymology, but it is remarkably serviceable as a definition for the "wild," that is, nature that has not been submitted to and, perhaps, even resists human control. In the Romantic period, the Arctic remained a largely undiscovered region where nature seemed "self-willed" in its frequent wrecking of ships and taking of lives. In this chapter, I will discuss the Arctic as an image of the "wild," focussing on the novella *The Surpassing Adventures of Allan Gordon* (henceforth cited as *SAAG*) by the Scottish writer James Hogg (1770–1835). This is the tale of the shipwrecked Allan Gordon, who spends several years in the high northern latitudes, much of the time alone and all of the time under threat of being eaten by hungry polar bears. The attempt to "unwild" the Arctic, to turn it into a safe human habitat, is at the heart of the story, which is loosely based on *The Life and Strange Surprizing Adventures of Robinson Crusoe* (1719).

The tale was published posthumously in 1837. Although the exact time of composition is not known, the story is clearly written in the wake of the Napoleonic Wars when naval resources were freed up and officers on half-pay could be re-employed to explore the Arctic. Spurred by the prolific whale fisher and explorer William Scoresby Jr.'s reports of relatively ice-free waters off Greenland's east coast in 1817, Sir John Barrow, the industrious Second Secretary to the Admiralty, launched a decades-long campaign to locate a sailing route to India and China either through the elusive Northwest Passage or across the North Pole. Hogg's story reflects on the ambition at the time that the Arctic could be subdued by man's knowledge, mapping, and control.

In the first part of the chapter, I will discuss Hogg's representation of the Arctic climate. *Surpassing Adventures* is written in a largely realistic register, drawing on the wealth of geographical and scientific information published in magazines and books in the wake of the reawakened enthusiasm for Arctic exploration. Hogg emphasizes the climate as essentially disorientating to the senses. Thus, the "wild" in the story is not only uncultivated nature in a physical sense, but also landscapes that frustrate our faculty for rational ordering. The disorientation of the senses presages Gordon's ultimate inability to bring the Arctic under human control. In the second part of the chapter, I aim to show that Hogg exposes man's desire to conquer the Arctic as a false "romance" that is replaced by a gritty narrative of hardship and final failure. *Surpassing Adventures* was meant as "a mere joke" on *Robinson Crusoe*, as Hogg explains in a letter to his publisher ("Letter" 209). However, this statement should not be read as an indication that the story is meant to be flippant. Rather, I suggest that the novella is a dark satire on Daniel Defoe's famous novel, questioning the colonial confidence of the precursor text. If Defoe shows us that the relations with the Other can teach an ego to master both the new land and the destiny of its inhabitants by his civilizing efforts, then Hogg's protagonist falls short on both counts. The narrative can be analyzed as existing in two parts. The first part focusses on Gordon's inability to comprehend, understand, and therefore ultimately conquer the Arctic. The second part centres on Gordon's integration with a floundering colony of white Europeans who compete for possession of the land with polar bears. I will contend that both parts of the story show human failure when pitted against the Arctic.

The Arctic wilderness

For the benefit of readers unfamiliar with *Surpassing Adventures*, it is useful to begin with a brief outline of the plot. Allan Gordon is a tailor's apprentice from Aberdeenshire. After a violent altercation with his hot-tempered master, Gordon decides to run away and go to sea as a cabin boy. In 1757, he finds himself on board a whaling ship, the *Ann Forbes*, bound for the Greenland Sea. The captain is a drunken and self-conceited Englishman, whose ambition is to discover the North Pole. Due to the captain's lack of judgement, the ship is caught in pack ice, which eventually crushes the hulk, leaving Gordon as the only survivor. The events related to Gordon drifting in either the Greenland Sea or the Arctic Ocean with the wreck of the ship stuck in a floe, to which an iceberg is attached, comprise a major part of the story. Gordon kills a large female polar bear that enters into the cabin where he is holed up. He learns that the bear had just given birth to a cub, which he decides to nurse. The cub, which he names "Nancy," grows up to provide Gordon with a surfeit of fish, thereby securing his survival, as well as defending him against other polar bears. Later, Gordon comes upon descendants of European colonists, who had come to Greenland from "Norgeway" (*SAAG* 298). The colonists take him to their diminished settlement in Old Greenland. Gordon is encouraged to take a wife as well as mistresses, licenced by the isolated colonists who are desperate for new blood. However, Nancy becomes jealous of Gordon's wife and eventually runs away. At the end of the story, Gordon leads a charge to kill the polar bears that regularly prey on the colonists, but this is forestalled by a legion of bears that launch a surprise attack on the settlement. However, one of the attacking bears turns out to be the runaway Nancy. She carries Gordon to safety before rejoining the feasting. Almost all the colonists are devoured, but Gordon manages to make his way back to Scotland on a Dutch whaling ship, which brings the story to its conclusion.

Hogg's text expresses disquietude about human ambition to master the Arctic and thereby challenges nineteenth-century British self-confidence about conquering the Arctic. This is most clearly expressed in the depiction of the English Captain John Hughes, who jeopardizes his ship and crew for the blind ambition of discovering the North Pole. It should be noted here that Hogg had previously satirized British determination to know the secrets of the poles. In an 1829 verse narrative, Hogg invents the larger-than-life character Jock M'Pherson, who tells us that he has crossed both the poles and seen "the sockets they ran in" as well as the giants who turn the earth on wheels oiled with "cans of bear's grease." Jock magnanimously advises the contemporary polar explorers, "Barrow, and Parry, and Franklin," to "commence" their explorations and "learn to speak sense" on the basis of his revelation of the state of affairs ("P and Q" 694).

The Greenland Sea, which sets the scene for the story of Allan Gordon, was known for its ice formations, which were a danger to human life. Hogg's relocating of his tale of survival from Defoe's warm Pacific island to this icy world raises the stakes and justifies the ostentatious claim that Gordon's experiences "surpass" Robinson Crusoe's. Defoe's novel is the archetypal story of British colonialism, in which the protagonist will confidently assert himself to be "King, or Emperor over the whole Country which I had Possession of" (151). In this respect, Gordon underperforms, unable to exert control over his new environment.

The Arctic and Antarctic were the locations for some well-known Romantic-period stories, most significantly S. T. Coleridge's "The Rime of the Ancient Mariner" (1798), Mary Shelley's *Frankenstein* (1818), and not least Edgar Allan Poe's *The Narrative of Arthur Gordon Pym of Nantucket* (1838), which also includes a shipwreck. These narratives show the Polar Regions as eerie and mystical climes where supernatural happenings take place and man may come undone. This is also an important aspect in the first part of *Surpassing Adventures*. After being shipwrecked off Spitsbergen (*SAAG* 248), Gordon's first-hand account details his psychological struggle to come to grips with the strangeness of the Arctic wild, as the climate disrupts his ability to rationally order his sense impressions. Gordon finds himself in a state of constant phenomenological uncertainty. He soon realizes that the high latitude of the Greenland Sea renders time out of joint, as he loses "all reckoning of months, weeks, or days," the only means of time-keeping being the length of his beard (*SAAG* 257). In the summer season, there is also the phenomenon of the midnight sun, which makes it impossible to tell "one time of the twenty four hours from another" (*SAAG* 266). In their work on the Arctic and the history of its exploration, Charles Officer and Jake Page note that in a place where time zones converge, time loses its meaning and space takes on a much more important dimension (3). But, for Gordon, even spatial orientation is bewildered, and he experiences strange illusions caused by atmospheric phenomena. For example, he often sees "hills and glens, with wreaths of snow here and there," but "could never see them in the same direction again" (*SAAG* 267). This inability to make determinate conclusions based on his faculties of sense makes him proclaim that the Arctic is "a strange unearthly climate" that "has no congeniality in it with human nature" (*SAAG* 267). It is often hard to decide whether observations are real or illusory. For example, Gordon sees a female figure on land who soon disappears, and no other humans are seen for weeks. He also hears a gunshot, but never discovers who could have fired a gun. For Gordon, the Arctic, with its refractions of images and sound, constitutes an epistemological wilderness, intoned with an almost biblical sense of a being spiritually lost in the desert.

Hogg himself never visited the Arctic, so his descriptions are based on published travel accounts. From his readings, he would have known the Arctic as a place where ordinary sense perception is suspended from the popular account published by the whaler and Arctic explorer William Scoresby Jr., who devoted several pages of his *Journal of a Voyage to the Northern Whale-Fishery: Including Researches and Discoveries on the Eastern Coast of West Greenland* (1823) to optical illusions and strange Fata Morgana that confuse and confound the explorer. The book included an illustration of a "grand phantasmagoria" of cities, ruins of castles, churches, monuments, etc., that appeared before the crew's eyes in the icy desert (160–173). Arctic mirages were further analyzed by the Scottish physicist Sir David Brewster in the entry on "Ice" for the *Edinburgh Encyclopædia* and, at greater length, in his *Letters on Natural Magic Addressed to Sir Walter Scott* (1830). In both texts, Brewster draws heavily on Scoresby's account of the Greenland Sea. It is important that Hogg's *Surpassing Adventures* is "Humbly and most respectfully inscribed" to Brewster, whose influence is also clearly discernible in the plot. Brewster's most famous theory was the existence of an Open Polar Sea (*Observations*). At one point, Gordon's ice floe drifts into this mythical ice-free water (which does not exist), before he is again carried south by a rapid current (*SAAG* 272–276). What attracted Hogg to the Arctic as a theatre for his story was that it was not only unknown and exotic, but also bewildering and disorientating to the senses. Thus, the modern condition of the instability of the self is expressed through the metaphors of meteorological chimeras. This is essentially what is at the heart of Hogg's best-known text, *The Private Memoirs and Confessions of a Justified Sinner* (1824), in which two differing accounts of the same events follow each other. This suspension of certainty is epitomized in the novel's scene at Arthur's Seat (61–62) where a hallucinatory doppelgänger is seen by means of a so-called Brocken Spectre, that is, a mountain phenomenon by which the shadow of the observer is cast upon clouds opposite the sun's direction.

Hogg had an enduring interest in optical illusions, and he described several such phenomena experienced in the wild and remote hills of Scotland. For instance, he experienced the "optical delusion" of seeing trees "flourishing abroad over the whole sky," when caught in the Scottish snow blizzard of 1794 ("Shepherd's Calendar" 165). An extended description of strange sightings is found in Hogg's autobiographical essay "Nature's Magic Lantern" (1833), which has examples of illusions observed in the "particular states of the atmosphere" of the remote Scottish hills (360). In the same work, Hogg recounts having seen a strange double shadow of himself, which he would later discuss with David Brewster, who accounted for it "by some law of dioptrical refraction" (356). So, in several ways, the interest in the Arctic is an expansion of Hogg's long-standing interest in communicating to the reading public a sense of the Scottish landscape as a spirited locus for folk beliefs in magic, fairies, and the supernatural. The bugbears of the many superstitious tales he integrated in his writing are here replaced by the mysterious polar bears. Folklore's fear of hostile powers is replicated in the scenes following the shipwreck. Gordon has managed to kill a huge she bear, and the forepart of her body is hanging out of the cabin window, yet he hears a strange munching sound. This makes him "more frightened than ever," and he begins to think "nature was all reversed in that horrible clime" (*SAAG* 259). However, as is often the case in Hogg's stories, this turns out to be the narrator's own mind playing tricks on him. It turns out that the sound comes from the bear's young cub gorging on her mother's spoils outside the cabin window.

For an analysis of the story's many examples of cognitive distortion, we would do well in turning to a theory that had much bearing at the time. The inability to mentally process sense phenomena is closely related to the category of "the sublime," as theorized by Edmund Burke in *A Philosophical Enquiry into the Origin of Our Ideas of the Sublime and Beautiful* (1757). Burke's famous thesis predated the intensified interest in the Arctic, and examples of the "sublime" from this bioregion are therefore notably absent in his writing. However, Immanuel Kant in *Critique of Judgement* (1790) mentions "shapeless mountain masses piled in wild disorder upon each other with their pyramids of ice" as a scenery of chaos that may bring about a sense of the sublime in the observer (117–118). The menacing landscapes of Arctic panoramas with their huge icebergs, massive glaciers, wrecked ships, and vistas of infinite icy landscapes found their way into a long line of late eighteenth- and early nineteenth-century European painting, *tableaux vivants*, and fiction, as has been discussed in studies by Robert G. David and Russel A. Potter. The "Arctic sublime" was an aesthetic vogue that became a category in its own right, which has been examined in a seminal article by Chauncey C. Loomis and has since been further qualified by Cian Duffy (102–124).

Hogg would have read Burke, whose examples of the sublime, when related to natural phenomena, are all from "wild" nature, not the cultivated garden. It is the danger of damage or death by such forces that will give delight, when experienced (or described) from a safe distance. To give an example, Caspar David Friedrich's famous oil painting *Das Eismeer* [The sea of ice] (1823/4), which is the spectacle of a shipwreck that lies immobile, quashed under huge shards of ice, is often hailed as the prime example of the "Arctic Sublime" (Potter 57–59; see Figure 11.1). Hogg's descriptions of the Arctic's power over the wrecked *Ann Forbes*, submerged in "broken ice towered up in heaps" (*SAAG* 248), point to a very similar spectacle.

One of the unequivocally sublime moments in *Surpassing Adventures* unfolds when Gordon observes the formation of new ice. Gordon is roused from sleep by a piercing sense of "undefinable terror" caused by "a rushing noise sound" (*SAAG* 277). This sound then modulates into "loud crashes, like discharges of artillery," after which he witnesses "a scene with which I shall never see any thing again to compare":

> The ice had thickened to a board, several inches in thickness; but the form of this huge mountain, with its broad shelving base running in below the ice with the current, heaved up the field into broad crystal flakes, which gradually rising to a perpendicular position to the height

Figure 11.1 "Das Eismeer" (1823–1824) by Caspar David Friedrich; Kunsthalle Hamburg, Hamburg (Wikimedia Commons).

> of an hundred feet and more . . . Again and again was the great frozen space broken up with crashes not to be equalled by any thing in nature, and therefore incomparable, unless we could conceive the rending of a sphere to pieces. (*SAAG* 278)

All the elements of a sublime experience are present in this passage. There is bewilderment, danger, and the dislocation of the rational understanding. The above passage recalls a scene described by Moravian missionaries in Labrador and cited in the entry on "Ice" from Brewster's *Edinburgh Encyclopædia*. This was a spectacle of "immense fields of ice" rising out of the ocean, "clashing against each other, and then plunging into the deep with a violence which no language can describe, and a noise like the discharge of a thousand cannon [*sic*] . . . a sight which must have struck the most unreflecting mind with solemn awe" (11: 638, rpt. from Brown 1: 603). Sublime spectacles like this were carefully curated in nineteenth-century publications, helping to inscribe the Arctic with a sense of wildness that poses a challenge to man's powers, both physically and mentally. To Burke, the sublime is an emotion that operates in a manner analogous to terror, which in its final instance is the fear of death.

One thing that characterizes Hogg's description of the Arctic is its excess of illumination. Burke had noted that light, when it overwhelms the sense of sight, can sometimes be elevated to the category of the sublime (144). An example of this is when Gordon finds himself engulfed in "rime," or frost fog, illuminated by light to the extent that it becomes a "dazzling whiteness" that totally blinds him (*SAAG* 286). The blanket of light poses a real danger, because it envelops him at a most critical moment when hungry polar bears pursue him. Gordon escapes with the help of Nancy, but the impaired vision is a portent of his ultimate inability to master his new environment. The consequence of his metaphoric blindness is the destruction of the human colony he promises to protect, handing final victory to the polar bears.

The Arctic: from romance to tragedy

Surpassing Adventures is framed as Robinsonade in the Arctic, and, as I will argue, Hogg is consciously using the framework of Defoe's novel to confront the myth of man's aptitude for holding dominion over nature. A central theme in *Robinson Crusoe* is the remaking of the natural environment, incrementally turning the foreign milieu into a liveable habitation that mirrors the structure of Crusoe's home country. In Jean-Jacques Rousseau's oft-quoted celebration of *Robinson Crusoe*, it was Defoe's focus on man's ability to "preserve [his] life," "find food," and procure "a certain amount of comfort" that made it "the best treatise on an education according to nature" (147). Indeed, since the Middle Ages, to tame nature has been seen as man's duty to God. Carolyn Merchant summarizes this idea: "Civilization is the final end, the telos, toward which 'wild nature' is destined." This is the "progressive narrative" that "undoes the declension of the Fall" (147). In *Surpassing Adventures*, Gordon manages to modify the Arctic, if only to a moderate degree. He carves out "apartments" in his iceberg, making it replete with a fire vent and a pantry. And it becomes Gordon's safe place, which he refers to as his "castle" (*SAAG* 268), echoing the term Crusoe uses to describe his abode. The ability to make a home out of an ice floe in the ocean is, of course, a remarkable feat; it is to inhabit what Siobhan Carroll has termed an "atopic space," that is, a region "deemed penetrable but inhospitable," such as the atmosphere, the oceans, and the poles (7).

In much of Hogg's work, there is an underlying tension between a conscious application of romance and anti-romance.[1] That is to say, he will sometimes simultaneously pander to and interrogate romantic values that he was often expected to deliver, engendering complex irony. In this respect, it is revealing for the interpretation of this part of Hogg's story that Gordon refers to the iceberg as his "romantic mountain" (*SAAG* 269), as if it were placed in a more temperate climate surrounded by green pastures. Gordon believes that he has a rapport with the landscape, illustrated by the information he relates of often shouting into a cavern, which will return an "exceedingly romantic voice" as if spoken by the "spirit of the iceberg" (*SAAG* 265). It is impossible not to read this romanticizing of an echo as emblematic of how Gordon is projecting an idea of romance onto the Arctic landscape. "Romantic" may therefore be read in what was a commonly used sense at the time: "characteristic of or befitting a romance" (*OED* 2a). Icebergs had long been objects of fascination and "formed the foundation of many a romantic tale of the middle ages," as the medieval historian Francis Palgrave noted in an 1818 review on Arctic travel. The "translucent and attractive islands" observed by nineteenth-century explorers "remind us at once of the mountain of adamant of Sinbad the Sailor, and of Huon of Bourdeaux, and of Duke Ernest of Bavaria" (493)—all of them well-known stories from medieval tradition with romance elements.

Gordon reflects on his time with Nancy on the iceberg as "one of such romance, that if I could have been certified that at any future period I should escape to give a relation of it, I would have chosen to remain for the present" (*SAAG* 291). As Gillian Hughes notes (23), Nancy resembles Friday, as can be seen in the reference to her as a "young savage of the desert" (*SAAG* 262). We may here remember Peter Hulme's observation that Crusoe's description of Friday is "tinged with erotic delight, though this is not easy to separate from a master's joy in a well-proportioned and healthy slave" (212). The same dynamic is present in the relationship between Gordon and Nancy. When Gordon first sees the grown female bear that is Nancy's mother, he compares her to a "naked woman . . . altogether without clothes" (*SAAG* 252). The association is presumably based on the fact that her silky, white coat is smooth like female skin (see Burke's comparison of "smooth coats of birds and beasts in animal beauties" to "fine women; smooth skins" [213]). Gordon names the she bear's cub after his first sweetheart (*SAAG* 262), and he grows so fond of Nancy that he will confess: "I loved her sincerely, I might almost say, intensely" (*SAAG* 264). When he grooms her, it is to make her "clean as a bride" (*SAAG* 288). Gordon and Nancy also sleep together every night (*SAAG* 289). This strange domestication of the icy world, by other writers often depicted as red in tooth and claw, takes a burlesque turn when the two promenade on the ice, as if they were a couple enjoying a stroll on a country road:

> I took her paw in my arm, and learned her to walk upright. A pretty couple certainly we were, I dressed like a gentleman in my late captain's holiday clothes, and she walking arm in arm with me, with her short steps, her long taper neck, and unfeasible, long head; there certainly never was any thing more ludicrous. (*SAAG* 265)

Gordon concedes, at this point, that imposing a civilizing yoke on a polar bear is a ludicrous proposition. Increasingly, the ability to fit Nancy into a human domain becomes ever more strained. It is symbolically significant that she eventually grows so big that her body nearly suffocates Gordon when they sleep in the same bed (*SAAG* 307). When Gordon decides to take a wife among the Norse colonists, Nancy runs away, finally realizing her incommensurability with human society. Gordon's attempt at weaning a polar bear from its natural instincts and making it his human companion will eventually be revealed as an unsustainable folly, a quixotic attempt to control the Arctic that is doomed to failure.

When the literary term "romance" is used in the novella, it is meant to ring a warning bell to the reader. It tells us that the respite from the hardships expected in the Arctic is a temporary evasion of reality. Not unexpectedly, the romance comes to an end when Gordon goes to live with the Greenland colonists, which marks the introduction of the reality principle in the story. Having lived in Greenland, isolated from the surrounding world for centuries, they are an image of the decidedly unromantic hardship and loss of civilized living that outsiders taking up habitation in the Arctic will inevitably suffer. Yet the colonists seem as deluded to this reality as Gordon was in the first half of the story. They believe their country to be "a terrestrial paradise," although Gordon judges the Greenland colony to be "the bleakest and last abode of living men" (*SAAG* 308).

The Greenland community is made up of white Europeans, the "remnant of a colony of Norwegians," who have survived in "Old Greenland" (*SAAG* 300). The reference here is historical and points to the Norse settlers who came to Greenland in AD 985 and cultivated the wild fjords in the south. At their peak, the settlements counted several thousands, but all communication with colonial Greenland was cut off at the beginning of the fifteenth century. Sailing routes were forgotten, and it was believed that the formation of pack ice off the east coast had hemmed in the colonists (in fact, settlements had been on the southwest coast). Today, we know that the Norse settlers abandoned Greenland around 1450, although the reasons are not yet fully understood. The survival of the colony was hypothetical, as it had not been visited for 400 years. Yet the possibility that their descendants would still be there was acknowledged by many in the early nineteenth century. Generally, it was held that the Greenland climate would prevent descendants of the original colonists from maintaining civilization. Hogg follows up on such suggestions, picturing them as living in destitution—a more realistic image of how Robinson Crusoes would cope without supplies from Europe.

Another claim regularly repeated in the early nineteenth century was that the Norse settlements had been raided by invading Inuit, who had entered Greenland from North America. The Greenland expert Charles Giesecke wrote for *The Edinburgh Encyclopaedia* that attacks took place when the Norse settlers were "weakened to such a degree that at last it became an easy matter for the Esquimaux (called Skraellingers by the settlers) to make war upon them and to extirpate them" (1). Hogg, whose tale concludes with the destruction of the colonists, avoids placing the blame for genocide on the Inuit and instead recasts it as an attack by polar bears. In *Surpassing Adventures*, the bears embody a combination of two regular features that literary historian Rebecca Weaver-Hightower has identified as endemic to the castaway-on-an-island genre: the confrontations with cannibals, who represent the ultimate lack of control for the castaway, and dangerous animals that "carry the island's infectious savagery" (94, 144). Usually, the castaways repel both of these threats, but not so in Hogg's story. As Sarah Moss notes in a short comment on *Surpassing Adventures*: "Hogg's Norse Greenlanders die, essentially, of polarisation, fearing and at last consumed by the frozen shapes that kill" (par. 17). The savageness of the Arctic is eventually reinstated with almost vindictive relish, as the "white bears" devour the colonists with "joy and triumph," emitting "growls of voluptuous joy" (*SAAG* 314).

Figure 11.2 "Man Proposes, God Disposes" (1864) by Edwin Landseer; reproduced by kind permission of Royal Holloway, University of London.

Hogg's scene of violence is unusually graphic and can be seen to anticipate Edwin Landseer's controversial oil-on-canvas *Man Proposes, God Disposes* (1864), which shows two polar bears brutally tearing at human remains in the polar wilderness (see Figure 11.2). Landseer's direct reference is to the fate of Sir John Franklin's failed 1845 expedition, from which the bones of the crew had been found to have been gnawed. Accusations that cannibalism had taken place were made, but Francis Leopold McClintock, who wrote a report on the incident, suggested that the expedition's remains were perhaps despoiled by polar bears (176). If this is the immediate reference for Landseer's gruesome vision, Diana Donald has suggested that the painter may have been influenced by Hogg's story (pars. 13–14). I may add here that Landseer may indeed have known Hogg's works, as Landseer made friends with Hogg's patron, Walter Scott, and one of his favourite subjects for painting was the shepherd's dog, for which Hogg's articles and stories were the most popular resources. What can be said with confidence is that the title of Landseer's painting (a version of Proverbs 19:21) resonates with Hogg's representation of man's helplessness in the Arctic. Like Crusoe, Gordon continually praises divine Providence for his survival (*SAAG* 260, 268, 271, 278, 282, 285). However, his ingenious proposal to defeat the polar bears by ambushing their habitat is mocked by the massacre on the colonists that is the result. The lesson is that God's Providence cannot be taken for granted, and, when man is at his most vulnerable in the wild, it may be withdrawn at any moment.

Hogg is writing his story as an intertext to *Robinson Crusoe*, because it allows his own bleak vision of nineteenth-century attempts to master the Arctic to stand out in relief against the confidence of Defoe's protagonist. If the general narrative drive in Defoe's novel is to let the forces of civilization slowly gain the upper hand over the "wild," then Hogg insists on the Arctic's final victory. To return to Thoreau's false etymology, wild nature is "self-willed." This also goes for the inhabitants who live in the wild. Whereas Crusoe successfully converts Friday from the abhorrent practice of cannibalism, Gordon fails to wean Nancy off eating human flesh. And if Crusoe successfully ambushes and kills a group of cannibals with the help of Friday, Gordon's surprise attack on the bears backfires, and he is outsmarted by the bears. The colonists are eaten, gleefully assisted by Gordon's erstwhile companion.

That Hogg, in this part of the story, wants the bears to fill in for Defoe's cannibals seems to have gone unnoticed in criticism despite the clear structural parallels. For example, Gordon discovers what he believes are human footprints in the snow, only to learn that they have been left by polar bears. This recalls the scene when Crusoe is surprised by a footprint, which he later realizes belongs to a cannibal. Gordon's planned attack on the polar bears finds a parallel in Crusoe's liberation of his fellow humans from the feast of cannibals in chapter 25, only that Gordon fails and everyone is eaten. The account of the final massacre describes a lustfully vampiric feasting: the bears "seized a number of the women," whom they "clasped in their paws, and then stretching them on the snow, they embraced them to death and sucked their blood" (*SAAG* 313). This appears to be an ironic compensation for Gordon having killed and skinned Nancy's mother, as a consequence of which the young cub was left licking her skin in vain, "looking for what it could not get, the mother's exhausted dug" (*SAAG* 261).

To be eaten was the prevailing fear of colonial discourse that symbolizes a wider and more diffuse anxiety about one's identity being consumed by the new environment. If *Surpassing Adventures* concludes with a sudden and brutal devastation of the Norse colony, the real tragedy is the slow and inevitable chipping away of European identity and civilization in the "wild." It is to this aspect of Hogg's story I will now turn.

The colony in Greenland

Much of Hogg's poetry suggests a value for the pastoral life in which he himself had grown up. The pastoral allows humans an integral place within nature and gives them stewardship over animals, as Hogg describes it warmly in his autobiographical memories about his days as a shepherd in the Scottish borderland, helped by his beloved collie dogs (for example, "Anecdotes of the Shepherd's Dog," "Further Anecdotes of the Shepherd's Dog" [1818], and *Shepherd's Calendar* [1819]). Joseph Meeker has defined the pastoral as a "halfway house . . . somewhere midway between the horrors of wilderness and the horrors of the city" (91). The European Greenlanders have definitely crossed the line into the "wild." When Gordon encounters the colony, it is in a sorry state. Having once counted thousands, there are only thirty-one men, ten women, and seven children left, as many have perished at sea or in bear hunting. But, most critically, the settlements are attacked by polar bears just before the time of their hibernation (*SAAG* 300–301) when the predators must gorge themselves to survive.

The colonists are a realistic vision of how these long-term "Crusoes" may cope after centuries of isolation and struggle for survival in the wild. The colonists still keep the Sabbath and pray, but all their prayers are for food and protection against polar bears. Furthermore, they have slackened their Christian morals by allowing every man to take "handmaidens" in order to increase the colony's dwindling numbers (*SAAG* 301–302), putting survival above moral strictures. The ignominy of having several partners beyond one's wife was a feature often discussed in relation to the Inuit in Greenland by the missionaries who recolonized Greenland from 1721.[2] The indication here, as in other descriptions given in the story, is that the Europeans have "gone native." Life in Old Greenland is at subsistence level with the "repetition" of endless hunting and fishing from "year to year" (*SAAG* 310). The European colonists have been reduced to the primitive life of the Eskimo by eating raw or even frozen fish—Gordon preferring the former (*SAAG* 307). This particular piece of information seems to build on the contemporary theory that the ethnonym "Esquimaux" is translated as "eaters of raw flesh" (O'Reilly 45, 50, 52). Hogg is here playing on the cultural anxiety of "going native," which articulates the fear of backsliding from civilization. A repository of scare stories about succumbing to the Other haunted the Western mind at the time of colonial expansion. In Hogg's tale, it is the Arctic itself, more precisely the ice barrier off the coast, that holds the colonists hostage. Greenland is the harshest of wildernesses where even the most fundamental Cartesian distinction between animal and man seems to be transgressed. The colonists have degenerated to the level of mere animal existence, and now practically hibernate during the winter months. As Hogg's protagonist concedes, his newfound community is "little better than the bears lying in a torpid state," becoming "drowsy, insipid, and almost incapable of moving" (*SAAG* 310).

In the section devoted to life in the Norse colony, Hogg answers some of the contemporary speculations about the fate of the Norse Greenlanders. To give just one example, there is the introduction to the 1818 reprint of the missionary Hans Egede's *Description of Greenland*, which contains the following series of questions:

> . . . have they perished by the inclemency of the climate, and the sterility of the soil? or do they still subsist? If they subsist, it must greatly interest our curiosity to learn in what manner they have vanquished the difficulties with which they have had to contend, both from the climate and the soil, and the total privation of all articles of European manufacture . . . Have they remained nearly stationary at the point of civilized existence at which their ancestors were placed four centuries ago? or have they entirely degenerated into a savage race, and preserved no memory nor vestige of their original extraction from, and subsequent communication with, the continent of civilized Europe? ("Introduction" xi–xii)

These questions address well-worn concerns about the loss of faith and identity when one came into contact with the Other, key among which was the question of whether European Christians were able to retain their identity over a considerable span of time when plunged into the wilderness. Frederick W. Turner notes that the mythos dramatized in tales of being lost in the wilderness is often a Christian fear of becoming possessed by the wild spirit, whether it is channelled by savage people or the "wild" itself (236). In other words, colonizers fail in their endeavour and are themselves colonized by the savagery of the foreign place. This stands in stark contrast to the usual solution adopted in the sub-genre that came to be known as the Robinsonade, in which the ordeal turns to triumph. The Robinsonade is often broadened to include the social life of a family or small community, which invites Utopian ideas towards which the genre is ideologically attuned. Hogg's story was given this more traditional direction when it was adapted for a juvenile audience by the American author Arthur Roth in *The Iceberg Hermit* (1974). In Roth's version, Gordon's time with the Greenlanders is depicted as a period of simple, honest, and near-Rousseauian living, while the concluding massacre is entirely elided.

Hogg wrote *Surpassing Adventures* at a time when the British Admiralty was determined to conquer the Arctic, part of which was the essentially "romantic" programme of reaching the North Pole. Hogg ends his Robinsonade with a foreboding of man's inevitable defeat, which would come to pass a few years later with the loss of the Franklin expedition. To end where this chapter started, with semantics, Hogg's tale is about survival in the "wilderness"—a word Samuel Johnson defined in *A Dictionary of the English Language* (1755) as "a tract of solitude and savageness." This is what Hogg imagines the Arctic to be like. But a rough summary of the actual etymology of "wilderness" points more towards savageness than loneliness, as it suggests a place where wild animals live. As Michael Engelhart points out in a book-length study on polar bears, this animal, with its white fur and carnivorous nature, has long been seen as the Arctic's most recognizable "icon." In Hogg's *Surpassing Adventures*, polar bears take on the role as *genii loci* of Greenland's hostile nature. It is to the bears, the true natives of the Arctic, not the human interlopers, that the "wild" belongs.

Notes

1 For Hogg's yoking together of romance with anti-romance (or the ironic), his 1822 novel *The Perils of Man* is often mentioned as a primary example. Gifford has analyzed how these oppositional modes are integrated in Hogg's writing (xiv–xxv).

2 For a description of concubinage and polygamy among the Inuit, see Crantz 1: 145–149.

Works cited

Brewster, David. "Observations on the Mean Temperature of the Globe." *Transactions of the Royal Society of Edinburgh*. W. and T. Tait, 1823, pp. 201–225.

Brown, William. *The History of the Propagation of Christianity among the Heathen since the Reformation*, vol. 1. A. Fullarton & Co., 1823.

Burke, Edmund. *A Philosophical Enquiry into the Origin of Our Ideas of the Sublime and Beautiful*. 4th ed. R. and J. Dodsley, 1764.

Carroll, Siobhan. *An Empire of Air and Water: Uncolonizable Space in the British Imagination, 1750–1850*. U of Pennsylvania P, 2015.

Crantz, David. *The History of Greenland: Including an Account of the Mission Carried on by the United Brethren in that Country*. rev. ed. Longman et al. 1820, 2 vols.

David, Robert G. *The Arctic in the British Imagination, 1818–1914*. Manchester UP, 2000.

Defoe, Daniel. *The Life and Strange Surprizing Adventures of Robinson Crusoe*. W. Taylor, 1719.

Donald, Diana. "The Arctic Fantasies of Edwin Landseer and Briton Riviere: Polar Bears, Wilderness and Notions of the Sublime," *Tate Papers*, no. 13 (2013), https://www.tate.org.uk/research/publications/tate-papers/13.

Duffy, Cian. *The Landscapes of the Sublime, 1700–1830*. Palgrave Macmillan, 2013.

Engelhard, Michael. *Ice Bear: The Cultural History of an Arctic Icon*. U of Washington P, 2016.

Giesecke, Charles. *Description of Greenland...From the Edinburgh Encyclopædia.* A. Balfour, 1816.
Gifford, Douglas. "Introduction," James Hogg, *The Three Perils Of Man.* Canongate Books, 1996, pp. vii–xl.
"*Greenland.*" *The Edinburgh Encyclopædia*, edited by David Brewster. Blackwood, 1830, pp. 481–502.
Hogg, James. "Letter to Blackie & Son, 25 March 1834." *The Collected Letters of James Hogg: Volume 4, 1832 to 1835*, edited by Gillian Hughes et al. Edinburgh UP, 2008, pp. 208–209.
———. "Nature's Magic Lantern." *Tales and Sketches*, vol. 4. Blackie & Son, 1837, pp. 352–366.
———. "The P and Q; or The Adventures of Jock M'Pherson." *Blackwood's Magazine*, vol. 26. October 1829, pp. 693–695.
———. *The Private Memoirs and Confessions of a Justified Sinner.* Longman et al. 1824.
———. "The Shepherd's Calendar." *Winter Evening Tales: Collected Among the Cottagers in the South of Scotland*, vol. 2, 2nd ed. Oliver & Boyd, 1821, pp. 152–204.
———. *Surpassing Adventures of Allan Gordon* (*SAAG*), *Tales and Sketches by the Ettrick Shepherd*, vol. 1. Blackie & Son, 1837, pp. 241–316.
Hughes, Gillian. "Reading and Inspiration: Some Sources for 'The Surpassing Adventures of Allan Gordon.'" *Scottish Literary Journal*, vol. 16, no. 1, 1989, pp. 21–34.
Hulme, Peter. *Colonial Encounters: Europe and the Native Caribbean 1492–1797*, new ed. Routledge, 1992.
"Ice," *The Edinburgh Encyclopædia; Conducted by David Brewster*, vol. 11. William Blackwood, 1830.
"Introduction." Hans Egede, *A Description of Greenland*, 2nd ed. T. & J. Allman et al. 1818, pp. xi–xii.
Johnson, Samuel. *A Journey to the Western Islands of Scotland.* W. Strahan and T. Cadell, 1775.
Kant, Immanuel. *Critique of Judgement.* Translated by J. H. Bernard, 2nd ed. Macmillan, 1914.
Landseer, Edwin. *Man Proposes, God Disposes.* Royal Holloway, University of London, 1864.
Loomis, Chauncey C. "The Arctic Sublime." *Nature and the Victorian Imagination*, edited by U. C. Knoepflmacher and G. B. Tennyson. U of California P, 1977, pp. 95–112.
McClintock, Francis Leopold. *A Narrative of the Discovery of the Fate of Sir John Franklin.* Murray, 1860.
Meeker, Joseph W. *The Comedy of Survival: Studies in Literary Ecology.* Scribner, 1974.
Merchant, Carolyn. "Reinventing Eden: Western Culture as Recovery Narrative." *Uncommon Ground: Rethinking the Human Place in Nature*, edited by William Cronon. W. W. Norton & Company, 1996, pp. 132–159.
Moss, Sarah. "Romanticism on Ice: Coleridge, Hogg and the Eighteenth-Century Missions to Greenland." *Romanticism on the Net*, vol. 45 (2007), pars. 1–17, https://id.erudit.org/iderudit/015816ar.
Officer, Charles and Jake Page. *A Fabulous Kingdom. The Exploration of the Arctic*, 2nd ed. Oxford UP, 2012.

O'Reilly, Bernard. *Greenland, the Adjacent Seas, and the North-West Passage to the Pacific Ocean*. James Eastburn, 1818.

[Palgrave, Francis], Review of Hans Egede Saabye, *Brudstykker af en Dagbok holden i Grönland, Quarterly Review*, January 1818, pp. 480–496.

Potter, Russell A. *Arctic Spectacles: The Frozen North in Visual Culture, 1818–1875*. U of Washington P, 2007.

"romantic, adj. and n." *OED Online*, Oxford UP, December 2019, www.oed.com/view/Entry/167122. Accessed 6 February 2020.

Roth, Arthtur. *The Iceberg Hermit*. Scholastic Inc., 1974.

Rousseau, Jean-Jacques. *Émile, Or, Education*. Translated by Barbara Foxley. J. M. Dent & Sons, 1961.

Scoresby, William. *Journal of a Voyage to the Northern Whale-Fishery: Including Researches and Discoveries on the Eastern Coast of West Greenland Made in the Summer of 1822*. A. Constable et al., 1823.

Turner, Frederick W. *Beyond Geography: The Western Spirit against the Wilderness*, rev. ed. Rutgers UP, 1992.

Turner, Jack. "The Wild and The Self." *The Rediscovery of the Wild*, edited by Peter H. Kahn and Patricia H. Hasbach. MIT P, 2013, pp. 27–50.

Weaver-Hightower, Rebecca. *Empire Islands: Castaways, Cannibals, and Fantasies of Conquest*. U of Minnesota P, 2007.

12 “Vast and irregular plains of ice”

Wilderness as smooth space in *Frankenstein*

Mirka Horová

> The wilderness has a mysterious tongue
> Which teaches awful doubt
>
> (P. B. Shelley, “Mont Blanc”)

> To move north . . . is to strip the self bare . . . to dispense with material and social consolations.
>
> (Peter Davidson, *Distance and Memory*)

The nineteenth-century imagination of Arctic wilderness is engendered by the infinitely fictionalizable blank space of the unknown. It is steeped in the spirit of epic discovery narratives, evoking the captivating volumes in Walton’s formative “uncle Thomas’ library” (9). As Janice Cavell explains, if “before 1818 the Arctic was still on the periphery of the Romantic imagination,” it gained prominence via “its parallels” to the Alpine sublime but also “its exoticism, distance, and difference” (305)—in other words, its radical otherness. According to Francis Spufford’s classic study *I May Be Some Time*, named after Captain Oates’s legendary last words, the “perception of Arctic emptiness” facilitated “wild . . . ideas of the region, for blank space, like blank paper, can be scribbled over with the wishes of the onlooker. There was a steady thread of fantasy concerning the poles in the nineteenth century” (83), and, as Adriana Craciun notes, even authentic accounts were often marked by “predisciplinary and generic fluidity” (86). The Arctic’s “unknown yet fiercely debated geography and climate” (Cavell 305) provide much more than a remarkable setting for Mary Shelley’s unsettling debut novel, however—wilderness is deeply interfused with its characters and ethos. The Arctic atopias, blank spaces on the map, represent one of the nineteenth century’s most challenging final frontiers. Their conspicuousness and projected economic promise naturally stirred steady imperial exploration, epitomized in the history of the so-called North-West Passage, fittingly described by Craciun as a “manifold entity” (30), a mutable construct. Originally “the untroubled settings of romance” and “myth,” as Siobhan Carroll argues, these regions “begin to appear vulnerable to the logic of state possession and control"—"mythical space is replaced by a blank space inviting penetration" (8). This Enlightenment desire "to penetrate the secrets of nature" (1831 ed., 45) is dissected in Shelley's novel with surgical precision, deftly staging the blurred line between exploration and exploitation.

Peter Davidson, citing Christopher Isherwood's 1938 novel *Lions and Shadows*, remarks on the trope of "the laborious, terrible North-West Passage" having become synonymous with "avoiding life" in "pointless exertion" (*North* 95). To some extent, over a century earlier, this conundrum already haunts *The Ancient Mariner*, the poem whose predatory imaginative vortex paradoxically sets Walton dreaming of polar delights, and Shelley's novel spins the egotistical twinning of empirical and poetic delusion in both Walton and Victor. While Frobisher and the long line of ensuing explorers represented a "national tradition of . . . sea-doggery" that the British were "eager" to live up to and outdo in the early nineteenth century (Spufford 54), and the North-West Passage represented "a character-defining challenge to the British Empire," Shelley's novel aims instead at imposing "limits" on this "imperial expansion" (Carroll 10). If, as Davidson observes, the "northward journey is . . . a journey into a kind of truth" (*North* 113), *Frankenstein*'s truth reveals the toxicity of the Enlightenment civilization project, exemplifying the flaws of Promethean science—or, in Jonathan Bate's assessment, the "Enlightenment proves to be endarkening" (51). Indeed, "instead of depicting the inevitable surrender of polar legend to the twin forces of imperial expansion and scientific inquiry" (Carroll 55), the novel explores in recurring detail the daunting power of Arctic wilderness, epitomized by the indefatigable agency of ice, and its substantial effects on the characters as well as the narrative structure. In a landmark move on Shelley's part, the Arctic ultimately resists "integration into the system itself" (Carroll 55), deconstructing the imperial ethos of discovery and destabilizing pragmatic conceptualizations of wild nature as something to be tamed and utilized. Equally, the novel problematizes certain aestheticizing impulses of the Romantic imagination of nature, critiquing their egotistical potential. It is this resistance of wilderness to both Enlightenment and Romantic egotism that is key in Shelley's novel, resonating deeply with the ecological challenges of our time.

Max Oelschlaeger's classic ecocritical study marks the nineteenth century's important shift "from viewing wild nature as merely a valuable resource (as a means to economic ends) and obstacle (wilderness must be conquered for civilization to advance) toward a conception of wilderness as an end in its own right" needing "preservation" (4). *Frankenstein*'s critique of the Enlightenment subverts "the belief that Western civilization represents the human triumph over an inhospitable environment" (4). The modern impulse "to relentlessly subjugate the wilderness, since things wild and free are alien to sensibilities nurtured so carefully in the garden of civilization" (8), is detrimental to humanity, as *Frankenstein* unflinchingly demonstrates.

In Edward Casey's recent phenomenological study of "edge-worlds," wilderness and wildness are differentiated as follows:

> Wilderness is a condition that holds for a given territory . . . shielded from cultivation and the inroads of human civilization. Wildness, by contrast, is a state of becoming that is not located in any particular place . . . [I]t is . . . in every place or region . . . indeed it is *in us*, in our stray thoughts, as well as in our unconscious mental life and repressed emotions. (138)

In *Frankenstein*, wildness and wilderness are intricately entwined in masterfully psychologized and geo-politicized *topoi* of transgression, involving Walton, Victor, and the Creature. In Michel Foucault's well-known concept, transgression "is an action that involves the limit" because it "incessantly crosses and recrosses a line," ever brought back "to the horizon of the uncrossable" (33–34). In the novel's manifold performance of the symbiosis of limit and transgression, crucial crossings and re-crossings occur in wild, uncharted space, still deemed eternal in the early nineteenth century—namely Alpine glaciers and the Arctic. This icy wilderness corresponds to Gilles Deleuze and Félix Guattari's concept of "smooth space"—an environment characterized by treacherous physical features, but also by the ways in which people perceive it, move in it, react to it, inhabit it. Smooth space is "occupied by intensities, wind and noise, forces, and sonorous and tactile qualities, as in the desert, steppe, or ice" (DG 479). Smooth space is "directional rather than dimensional," "filled by events or haecceities"—because it is a "space of affects" rather than "properties," "perception in it is based on symptoms and evaluations rather than measures" (DG 479). Striated space has been charted, systematized, encoded, while smooth space resists systematization and codification.

The tension between smooth and striated space in the novel is epitomized by the clash between the nomadic Creature who can subsist in the wilderness of smooth space, navigating it instinctively, if also by social necessity, and Frankenstein's and Walton's symptomatically striated modes of observation and interpretation of their unfamiliar surroundings. When faced with wilderness, they resort to the increasingly challenged traditional discourses of Enlightenment positivism or the Romantic sublime, only to find themselves epistemologically as well as literally at sea. While some recent studies have pointed out the correlation between Shelley's novel and Deleuze and Guattari's concept of the nomadic on textual, conceptual, and discursive levels (Heymans 118, 130), the novel's dynamics of smooth and striated space and the Creature's essential nomadism, which pertinently lends itself to interpretive exploration of the Romantic wilderness, have not been explored.

In cartographic terms, *Frankenstein*'s Arctic wilderness is the politically and ecocritically charged *terra nullius*—no man's land—uncharted, hostile, and unknown. In the novel, the Creature can only be approached by Frankenstein or Walton in these wilderness settings and on his own terms—on the Alpine glacier and in the finale among the Arctic ice floes. Walton's original efforts to striate the Arctic are thwarted, and, ironically, the only remaining contender in the forlorn polar quest and the discovery of a navigable Northern passage is the Creature, seeking "the most northern extremity of the globe" (161) in order to die there. His final escape on the ice raft confirms his formidable deftness at moving in the unnavigable, ever-changing wilderness of Arctic icescape. The Creature, as we shall see, represents the nomadic force, a swift vector—a force with direction, epitomizing the primacy of line and movement in smooth space, as opposed to the primacy of point and settlement in striated space. By necessity, the Creature only subsists in smooth space, but he also performs the nomadic in narrative terms, leading us deep into uncharted wilderness, geographically, conceptually, and textually.

Smooth space is characterized by sounds, events, and movement: "there is an extraordinarily fine topology that relies not on points or objects but rather . . . on sets of relations" such as "winds, undulations of snow" and "the creaking of ice" (DG 382). As Deleuze and Guattari suggest, "there is no intermediate distance, no perspective or contour; visibility is limited" (382). According to Barry Lopez's captivating classic *Arctic Dreams*, which neatly combines contemporary and historic exploration experience, observations of polar ecosystems, and the aesthetics of Far-North phenomena, the "monotony of the Arctic creates frequent problems with scale and depth perception" (239). Deleuze and Guattari's theory of smooth space explains that these problems arise because "there is no line separating earth and sky" (DG 382), something we would expect in striated space "canopied by the sky. . . and by the measurable visual qualities deriving from it" (DG 479). The following account of the 1786 Danish Löwenörn and Egede expedition to rediscover "the eastern coast of Greenland" showcases the Arctic's visual opacity:

> instead of land, they saw nothing but an immense and impenetrable extent of ice, in which rose enormous mountains of ice. The vapour and fog occasioned by the great extent of ice prevented them from seeing the coast, even if near it . . . The clouds assumed at times the appearance of land. (Barrow 330, 333)

Arctic wilderness is a "primal and atavistic landscape" (Fulford, Lee, Kitson 169), frustrating perception trained in striated space which is "defined by the requirement of long-distance vision," "constancy of orientation," and the "constitution of a central perspective" (DG 494). Through this striated lens, the "vision of the Arctic" persists "as bleak, blank, [and] hostile" (Spufford 82). As Walton remarks, citing Victor, "many things will appear possible in these wild and mysterious regions which would provoke the laughter of those unacquainted with the ever-varied powers of nature" (1831 ed., 37). Once encountered, the "physical landscape" of smooth space seems "wild and mysterious" because it is "an unstructured abode of space and time . . . not entirely fathomable" (Lopez 257). Furthermore, the Arctic's "real and temporal borders" subvert "conventional" time (Lopez 126), as Walton's letter from Archangelsk exemplifies: "How slowly the time passes here, encompassed as I am by frost and snow!" (10). This recurs in Victor's narrative, too: "I cannot guess how many days have passed . . . By the quantity of provision . . . consumed, I should guess that I had passed three weeks" (149).

Alongside unreliable visual impressions, the wilderness of Alpine glaciers shares with the Arctic the primacy of sound and movement. Sonorous "much more than visual," its dynamic "variability" and "the polyvocality of directions [are] an essential feature of smooth spaces" (DG 382). The glacier's sublime otherness is well captured in William Coxe's 1802 *Travels in Switzerland*: "I can no otherwise convey to you an image of this body of ice, broken into irregular ridges and deep chasms than by comparing it to waves instantaneously frozen in the midst of a violent storm" (II: 109). The Shelleys' *Six Weeks' Tour*, based on their 1814 and 1816 Alpine adventures, provides vital smooth-space characteristics of movement, sound, and stunning but treacherous visuals:

> our guides desired us to pass quickly, for it is said sometimes the least sound will accelerate [the] descent [of 'vast stones'] . . . Lines of dazzling ice . . . occupy [the summits'] perpendicular rifts . . . [and] pierce the clouds like things not belonging to this earth . . . In these regions, everything changes, and is in motion . . . The echo of rocks, or of the ice and snow which fall from their overhanging precipices . . . scarcely ceases for one moment. (165–167)

Smooth space is governed by extreme mutability—alertness and swiftness are key to navigating it successfully—something only the Creature is capable of, as we shall see later.

Victor's musings on the Alpine sublime offer further insight into the otherness of wilderness. As Cian Duffy observes, the Alpine "spatial vertigo" (128) is key here. Victor faces "the abrupt sides of vast mountains"; "the icy wall of the glacier overhung [him]"; "the ascent is precipitous," the mountain "perpendicular" (66). More importantly still, this "terrifically desolate" scene (66) presents a landscape perpetually and perceptibly in flux, governed solely by the principle of mutability. In this section, Mary cites Percy's eponymous poem, which reflects Lucretian materialism, highlighting wilderness as representative of nature in a radical state of becoming. In imagery recalling the *Six Weeks' Tour*, Victor marvels at "the solemn silence of this glorious presence-chamber of imperial Nature, "broken only by the brawling waves or the fall of some vast fragment, the thunder sound of the avalanche or the cracking, reverberated along the mountains, of the accumulated ice, which, through the silent working of immutable laws, was ever and anon rent and torn" (1831 ed., 90).

Victor's alertness to the mutability of smooth space is striking here. "The awful majesty" (67) of Mont Blanc does not teach "awful doubt" at this point, however—that is reserved for the Creature and the Arctic wilderness.

Victor praises the glacier's "absence of human thought" (Duffy 131) in a restorative, if also escapist, contemplation. The perilous wilderness of the glacier is "stupendous," transforming Victor's "sorrowful" "heart" "into something like joy" (67). Victor's vantage point, "overlooking the sea of ice," also shows his striated manner of observation, the visual aided by cause-and-effect descriptions. His impulse is to chart the smooth space of wilderness, both empirically and poetically, in his diction of the sublime.

The extreme otherness and mutability of smooth-space wilderness is well documented in *Frankenstein*—in Victor's descriptions of the Alpine sublime discussed above, the comparatively "desolate and appalling," "monotonous yet ever-changing" Orkneys (117), and the polar pursuit, complemented by Walton's deliberations on the Arctic. Here, otherness and mutability manifest themselves most palpably in the "secret ministry" of ice, to borrow Coleridge's magic phrase, which increasingly frustrates both protagonists. At first Walton boasts that "success shall crown [his] endeavours . . . the very stars . . . witnesses . . . of [his] triumph . . . over the untamed yet obedient element," hubristically asking: "What can stop the determined heart and resolved will of man?" (1831 ed., 32). The agency of ice, however, invariably prevails, deciding all: "we heard the ground sea, and before night the ice broke and freed our ship" (14). Similarly, for Victor, who wishes the ice to freeze rather than break at this stage, "immense and rugged mountains of ice often barred up [his] passage, and [he] often heard the thunder of the ground sea, which threatened [his] destruction. But again the frost came and made the paths of the sea secure" (149). These and numerous other accounts show both Victor and Walton at the mercy of the elements, since their methods of observation are too dependent on the visual; they only register the sound of the pack ice as a secondary, not decisive, factor of navigability. As Lopez explains, a "Western traveller in the Arctic . . . is inclined . . . to favour visual information over the testimony of . . . other senses" (271). Walton's account of first seeing a traveller in the distance (later identified as the Creature) exemplifies this:

> This appearance excited our unqualified wonder. We were, as we believed, many hundred miles from any land; but this apparition seemed to denote that it was not, in reality, so distant as we had supposed. Shut in, however, by ice, it was impossible to follow his track, which we had observed with the greatest attention. (13)

As Lopez explains, "because you have seen something does not mean you can explain it . . . [I]ndisputable information is a dot in space; interpretations grow out of the desire to make this point a line, to give it a direction" (127). This proclivity to visual orientation and imposed causality proves detrimental to both Walton and Victor, because smooth space does not conform to striated means of operation in Shelley's novel, underscoring its politics of wilderness.

To better grasp the significance of the Arctic's mutability in *Frankenstein* and its far-reaching empirical, philosophical, and political implications, we must highlight some key aspects problematizing polar exploration. The agency of polar ice is manifold, its effects bewildering. As Davidson notes, its "unidentifiability" features in "many accounts," "as if the imagination simply refused the concept of so much ice" (*North* 57). Lopez explains further that it is hard to imagine "any landscape more exhausting or humbling" (212). No wonder nineteenth-century accounts invariably conceptualized the Arctic as "war with the elements in a treacherous landscape . . . beyond the pale of civilization, a beast that preyed on virtue and enterprise" (Lopez 226). Positing the explorers' "virtue and enterprise" against an uncivilized-hence-beastly environment betrays features of Enlightenment prejudice. As exemplary smooth space, the Arctic is "a place where divisions do not hold," "the phenomenology of dualism cannot function" (Bate 54). Walton's and Victor's fatal flaw is summed up in their inflated Romantic idealism laced with imperial hubris, both overtly invested in the ego while misinterpreting the mutability of their surroundings. This is best seen in Victor's morale-boosting speech to Walton's mutinous crew: "This ice is not made of such stuff as your hearts may be; it is mutable and cannot withstand you if you say that it shall not" (155). In attempting to rouse the sailors with the vision of themselves as homecoming "heroes who have fought and conquered" (155), he utterly misunderstands "mutable" as weak and pliable rather than omnipotent and unpredictable. Walton's ship is, after all, "surrounded by mountains of ice which admit of no escape and threaten every moment to crush" it (153). In the novel's masterful portrayal of the ego-crushing mutability of smooth space, only inanimate nature (and the Creature, its complex symbolic denizen) retains agency, determining Victor's and Walton's fate by thwarting Victor's pursuit, partially conveying him to Walton's rescue, and finally freeing the ship:

> the ice began to move, and roarings like thunder were heard at a distance as the islands split and cracked in every direction. We were in the most imminent peril, but . . . we could only remain passive . . . The ice cracked behind us . . . a breeze sprang from the west, and . . . the passage towards the south became perfectly free. (156)

Arctic wilderness strips Walton and his crew of agency, rendering these explorers "passive" and powerless. As Lopez discerns, "pack ice moves irregularly"—it is "unpredictable" (214). Smooth space is not impossible to travel in, as the Creature's more successful and better adapted progress suggests, but it becomes so for both Victor and Walton as they struggle to tackle adverse conditions arising from the extreme otherness of the ice-governed landscape, owing to their reliance on striated methods of observation and navigation. While the Creature moves and subsists as a nomad, Walton's ship represents imperial enterprise, a vehicle of striation, maladjusted to the polar smooth space, and Victor's sledge pursuit is marred by the unpredictable agency of ice as well as his weakened state:

> when I appeared almost within grasp of my foe . . . I lost all trace of him more utterly than . . . ever . . . before. A ground sea was heard; the thunder of its progress, as the waters rolled and swelled beneath me . . . I pressed on, but in vain. The wind arose; the sea roared; and, as with the mighty shock of an earthquake, it split and cracked with a tremendous and overwhelming sound . . . [I]n a few minutes a tumultuous sea rolled between me and my enemy, and I was left drifting on a scattered piece of ice that was continually lessening . . . preparing for me a hideous death. (150)

Navigating pack ice feels like "walking over the back of some enormous and methodical beast" (Lopez 215). The Arctic's lesson of smooth-space mutability is ignored by the preoccupied Victor, once so attuned to the Alpine sublime, and missed by the reluctant Walton. Exploration-bound, he fails to grasp the introspective potential of the wilderness encounter. While Victor laments that his punitive "task is unfulfilled" (150), Walton bemoans his "cowardice and indecision," having lost "hopes of utility and glory," and declares: "I come back ignorant and disappointed" (155). The wisdom of the wilderness is, however, rather ingeniously reflected by the Creature and his planned suicide on the coveted North Pole, further subverting Enlightenment enterprise. As Eric Wilson notes of the Ancient Mariner in his ground-breaking *Spiritual History of Ice*, "the gnosis of ice is not a revelation of static truth"; "an insight into the mystery of ice re-releases the adept . . . wiser . . . but no closer to a stable principle in which he can rest" (219). While the Mariner lives to teach his bewildering tale for the supposed betterment of humanity, the Creature's "long and strange" (69) story leaves him and the reader at sea. According to Andrew Griffin, his resolution to "ascend [his] funeral pile triumphantly and exult in the agony of the torturing flames," seeking oblivion (161), can be read as "a bitter parody of both Walton's and Frankenstein's [Promethean] dream," thereby "underscoring the sorrow and fatality" of "the Romantic vision of being" (69–70). However, the Creature alone has grasped "the ice" as "a revelation of the groundless ground," setting off to become "undone, distributed in the all" (Wilson 219), though in a much darker sense than Wilson's Mariner. The Creature's final words reveal his deeply subversive humanity which is steeped in his classic reading and crystallized by Victor's death but contemplated during his life in the wilderness, when he is intensely attuned to its mutability: "The light of that conflagration will fade away; my ashes will be swept into the sea by the winds. My spirit will sleep in peace, or if it thinks, it will not surely think thus" (161). As Wilson concludes, in the wilderness' moral, "the ego melts, sinks back into the muck from which it arose. Yet, from this dissolution arise . . . new patterns of being that are less egocentric and more open" (220).

Wilson's meditative finale recalls the transformative potential of traditional religious wilderness encounters, edifying in dissolution, empowering in humility. While smooth-space wilderness absorbs the Creature, the waves bearing him away into "darkness and distance" (161), his fate and its moral remain unresolved.

Wilson's point about Victor requires more attention in terms of smooth-space dynamics: "That Victor dies in the Arctic ice is fitting, for he has existed (not lived) in an undifferentiated flatland ever since he shocked a cold body into motion. The ice is simply an externalization of his static interiors" (172–173). This reading is problematic because Wilson's own captivating "gnosis of ice" disproves anything "static." Besides, the novel's descriptions of the Arctic contest this notion, as we have seen. However, we may transpose Wilson's parallel of the "undifferentiated flatland" to smooth space in narrative terms: by abandoning the Creature, Victor effectively condemns him to perpetual exile. When the Creature confronts Victor on the glacier, the nomadic starts to dominate Victor's remaining story, bound to the Creature's. Similarly, Walton, another Enlightenment man trapped in unnavigable smooth-space wilderness, becomes a vehicle of multiple narratives and timelines, breaking his epistolary frame in increasingly unnavigable directions, highlighting the nomadic force of Shelley's complex narrative.

In order to grasp the key fault lines between smooth and striated space in *Frankenstein*, we must examine the conceptual tension of imagining, perceiving, and utilizing wilderness. The conundrum of smooth space is well summarized in Melanie Lörke's study of Romantic liminal phenomena:

> Smooth space is an ambiguous concept in Romantic texts. There seems to be a desire for smooth space to function as a medium that enables the loss of the self, that allows the subject to remain on the mountain forever, to disappear into the dark forest, or to sink into the stormy sea. But this desire also invokes an anxiety that calls for boundaries and the striated space of civilization and expresses itself through the search for the next village, the abandonment of the forest, and the avoidance . . . of the sea. (267)

In the case of Victor and Walton, the pull towards smooth space is marked by their quest for scientific discovery—signifying "the period's heady sense that the powers of nature might be appropriated for humanity" (Spufford 84). Even as they yearn for the bewildering unknown, to "pursue" the "hiding places" of nature (33), to "tread a land never before imprinted by the foot of man" (7), the impulse is to dis-cover—in other words, to striate. The Romantic desire for the wilderness' mediation of mystery twinned with the Enlightenment desire to explicate it is best summarized by Walton's first letter, where he shares his "ardent curiosity" and fantasizes about rendering the heavens' "seeming eccentricities consistent for ever" (7). The rationally posited "frost and desolation" of the North Pole is suppressed by Walton's "imagination," which spins instead the "region of beauty and delight" (7). Even the source of magnetism's "wondrous power" seems within reach—"if only Walton had known that the Magnetic North Pole itself" (Bate 55) migrates dramatically eastwards, from Canada to Siberia. While Walton's "trembling sensation" (12) is a Romantic commonplace, his polar dream "is a pure book-learnt construction of the imagination," "a space cleared on the map for him to fill with daydreams of discovery" (Spufford 84), betraying a fanatic strain of exploration-turned-exploitation, most apparent in the 1831 version: "how gladly I would sacrifice my fortune, my existence, my every hope, to the furtherance of my enterprise. One man's life or death were but a small price to pay . . . for the dominion I should acquire and transmit over the elemental foes of our race" (36). Combining the Enlightenment desire for "dominion" over "the elemental foes of our race" with an overdose of Romantic idealism—after all, Walton is a self-professed failed Romantic poet—is something Walton shares with Victor, whose captivating reflections on the Alpine sublime signify both the Romantic homage to nature and its escapist streak, once the mountains become a post-traumatic refuge: "Six years had passed since then: *I* was a wreck, but nought had changed in those savage and enduring scenes" (1831 ed., 88). Alpine wilderness recurs as sublime consolation—but, crucially, only until the fateful meeting of Victor and his Creature on the glacier. Shelley's use of wilderness is riddled with irony. While Frankenstein seeks solace in the sublime scenes of the Alps, "to forget [him]self and [his] ephemeral, because human, sorrows" (1831 ed., 88), he is confronted with the very thing he seeks to forget—his creation. Wilderness will manifest its own. At this key turning point in the novel, the Creature appears on the Mer de Glace as its emissary. If *Frankenstein*'s "wilderness has a mysterious tongue," to borrow Percy's line, its "awful doubt" is henceforth taught by the Creature.

As Paul Cantor observes, Victor's exile "with his creature to the Arctic wastelands becomes an emblem of the dangerous solipsistic tendency inherent in the Romantic concept of the imagination" (124). If the novel offers "a more complicated view of polar exploration . . . quintessentially representative of human ambition at its most grasping," in John McCannon's opinion (132), wilderness ultimately defeats Romantic idealism and Enlightenment empiricism, offering the reader introspection rather than expedition, leaving the explorers baffled and humbled. This introspection breaks new ground, leading to an appreciation of nature beyond the aesthetics of the sublime and the epistemology of discovery, whose flaws foreshadow the urgent need for an ethics of coexistence. Equally, Walton's seemingly innocent "favourite dream" of his "early years" (8) unravels in darker auguries of the Enlightenment's exclusionist and exploitative potential. While "conquer[ing] the Pole is presented as no less a Promethean feat than the creation of life" (McCannon 132), the hubris of the Western explorer betrays the Enlightenment's colonial prejudice, manifest in Walton's remark on rescuing Victor in the Arctic: "He was not, as the other traveller seemed to be, a savage inhabitant of some undiscovered island, but an European" (14). Victor's heroic speech to Walton's reluctant crew, steeped in imperial rhetoric, further heralds the threat that striation poses to the Arctic: the "expedition" was "an honourable undertaking" because it incited "courage" in the face of "danger and death," risking lives for "the benefit of mankind" (154–155). Victor's conceptualization of wilderness here is troublesome at best, especially considering the climate and extinction crisis that the last two centuries of pursuing "mankind's benefit" have wreaked on the Arctic and wilderness at large.

Frankenstein's subversive critique of Romantic idealism and Enlightenment endeavour complements its conceptualization of wilderness not as a frontier to be conquered but rather as smooth space capable of withstanding striation. Victor's famous dictum, "with unrelaxed and breathless eagerness, I pursued nature to her hiding places" (33), is key to our understanding of the novel's evocation and conservation of wilderness. The deftly eroticized sense of scientific pursuit—Victor and Walton's shared hunger for conquest—is defaced in Shelley's deconstruction of such masculine missions. Wilderness, a feminized noun, from the Germanic "land of wild beasts," comprises the "hiding places" of nature. The verb "bewilder" is semantically related to wilderness in crucial ways; wilderness leads astray and confuses. If, in the well-worn critical idiom, the Enlightenment sought to unravel and classify the mysteries of nature, and the Creature, a product of this new scientific endeavour, ironically becomes the embodiment of the mystery of wilderness, he also represents man-made, technological creation devoid of ethical consideration. Shelley's Creature interrogates humanity by signalling the posthuman. While the Creature is recognized as stronger and more resilient, surviving in adverse conditions, he is "forced to accept the civilized world's view of him as inferior" (Cantor 126) because he is considered sub-human. He recognizes his own physical superiority and uniqueness, however: "I was not even of the same nature as man. I was more agile . . . and could subsist upon coarser diet; I bore the extremes of heat and cold with less injury to my frame; my stature far exceeded theirs . . . I saw and heard of none like me" (83). Ostracized by the so-called civilized world for his appearance, the Creature embraces wilderness by necessity, and his nomadic experience provides further insight into the geo-political, philosophical, and epistemological dynamics of smooth space.

Temporary "occupation of atopic space" "is usually associated with mobile peoples—explorers, exiles, refugees, bandits, and mutineers—who have no place in, or have been dis-placed from . . . the nation" (Carroll 6–7), that is, from the striated space of civilization. As the Creature remarks: "I was dependent on none and related to none . . . The path of my departure was free" (89). In this respect, the Creature is less nomad than migrant, for if the nomad is "engendered in a series of local operations of varying orientations: desert, steppe, ice, sea" (DG 382), the migrant is not predicated upon locality, just forced movement. However, the Creature's own narrative provides us with material on his acquired nomadic existence which deeply resonates with Thomas Nail's characterization below:

> The nomads strive for a dynamic transformation of themselves according to the relative *pressures of the environment*. Without the possibility of sedentarism or the surplus of stored food, the nomads must continually launch and relaunch themselves back into movement. There is no stable ground to rely on, so the nomad must become its own ground in a continual oscillation of rigorous self-transformation and self-cultivation of pressure and counterpressure. (134)

Symbolically, the Creature's first place of rest is a "forest near Ingolstadt" (70); later he is "satisfied . . . with berries, nuts, and roots . . . gathered from a neighbouring wood," "not "heed[ing] the bleakness of the weather," being "better fitted . . . for the endurance of cold than heat" (77, 92). Abhorred and spurned by all who see him, starting with his creator, he finds "refuge" in "the desert mountains and dreary glaciers": "the caves of ice, which I only do not fear, are a dwelling to me . . . These bleak skies . . . are kinder to me than your fellow-beings" (68–69). Ultimately, "impassive" to "the misery of cold and frost," the Creature "seek[s] the everlasting ices of the north" (147). His "symbolically charged affinity with the realms of ice" (Cavell 301) is striking, conveying the novel's keen sense of smooth-space dynamics in relation to the nomadic.

The Creature's plan to "quit the neighbourhood of man and dwell . . . in the most savage of places," "the vast wilds of South America" (102–103), is key here. "[His] food is not that of man," as he notes; "acorns and berries afford [him] sufficient nourishment" (103). He proposes a simple gatherer's life with his partner-to-be-made, assuming that in the egalitarian wilderness, "the sun will shine on [them] as on man and will ripen [their] food" (103). Crucially, "it is not 'the hidden laws of nature' that interest" the Creature, but the outward "mechanics of nature and its . . . uses. To him the world is not 'a secret' to be . . . penetrated, but a place in which to live as comfortably as possible" (Griffin 66). Such an existence, as the Creature pointedly suggests, is "peaceful and human" (103), but also ascetically attuned to wilderness, something modern humanity has left behind. Our present-day concern at this stage remains that, should he "find an unoccupied swath of land to colonize, blank space will not remain remote" forever (Carroll 54). The pressures of striation are well observed here; however, the word "colonize" jars, as the Creature's plan is anything but that. Rather, the Creature articulates proto-ecological concerns in his radical sustainable vision of subsistence in wilderness.

Clairval's and Victor's leisurely trip along the Rhine and across Britain stands in sharp contrast to the Creature's experience of wilderness. Theirs is a comfortable journey through civilized nature—for example, Windsor's "beautiful forest" with its "stately deer" (114)—interspersed with Romantic poetry and Victor's reminiscences of his Alpine homeland once they reach Cumberland and Westmorland: "[he] could now almost fancy [him]self among the Swiss mountains" in those "familiar and dear sights" (115). This is a "pleasurable" rambling through nature, delaying the Creature's request, whose parallel pursuit is one of "toil and misery" (120). Mary Shelley does not romanticize wilderness: "I crept along the shores of the Rhine, among its willow islands and over the summits of its hills. I have dwelt many months in the heaths of England and among the deserts of Scotland. I have endured incalculable fatigue, and cold, and hunger" (120). In striated space, "lines or trajectories tend to be subordinated to points: one goes from one point to another," as in Victor and Clairval's journey. "In the smooth, it is the opposite: the points are subordinated to the trajectory" (DG 528), akin to the Creature's parallel pursuit. Locality and aim aside, movement and endurance are key in further differentiating the striated and the smooth.

The Creature's mode of movement is radically other—faster, more agile, and efficient. This characteristic features abundantly in the novel, contrasting the comparative slowness of others. While Victor "spent nearly two hours" crossing the glacier, the Creature advances "with superhuman speed": "He bounded over the crevices in the ice, among which I had walked with caution" (67). As Victor portends, "a creature who could exist in the ice caves of the glaciers and hide himself from pursuit among the ridges of inaccessible precipices was a being possessing faculties it would be vain to cope with" (104). On the glacier, Victor's crucial utterance marks the turning point in the narrative when the nomadic takes over: "he led the way across the ice: I followed" (69). Smooth space serves as mediation ground—it is a space of narrative negotiation. The Creature's and Victor's first conversation on the glacier represents a key crossroads, further branching the narrative until it becomes a nomadic experience, bewildering the reader's impulse to construct a linear story.

Once unleashed, the nomadic force epitomized by the Creature gains momentum, overtaking the narrative. The Other, be it wilderness such as the Arctic or the Creature, is branded bestial. The uncanny proximity of the wilderness to the Creature is often remarked upon: "Who can follow an animal which can traverse the sea of ice and inhabit caves and dens where no man would venture to intrude?" (144). While enacting justice, the pursuit of the Creature after the murders also betrays the Enlightenment logic of excluding otherness. At the deposition, Victor confides in the magistrate: "if he has indeed taken refuge in the Alps, he may be hunted like the chamois and destroyed as a beast of prey" (144). His wildness dominates Victor's accounts: "the strange nature of the animal would elude all pursuit . . . Who could arrest a creature capable of scaling the overhanging sides of Mont Salêve?" (51). Notably, the Creature himself stresses his initial restraint, alongside his rampant animality: "I could have torn [Felix] limb from limb, as the lion rends the antelope . . . I gave vent to my anguish in fearful howlings. I was like a wild beast that had broken the toils, destroying the objects that obstructed me and ranging through the wood with a stag-like swiftness" (94–95).

The elemental swiftness of the Creature disappearing into smooth-space wilderness is key here—apparent, for instance, in Victor's failed pursuit following the murder of Elizabeth: "I rushed towards the window . . . but he eluded me, leaped from his station, and running with the swiftness of lightning, plunged into the lake" (141). His proto-cinematic escapes feature extraordinary speed and stealth, framed by the pursuing eye: "his boat, which shot across the waters with an arrowy swiftness . . . was soon lost amidst the waves" (121). Similarly, on the glacier: "I saw him descend the mountain with greater speed than the flight of an eagle, and quickly lost among the undulations of the sea of ice" (104). Crucially, the novel's very finale is reserved for the Creature's last escape into wilderness: "He sprang from the cabin-window . . . upon the ice raft . . . He was soon borne away by the waves and lost in darkness and distance" (161). The creature is perceived as the ultimate nomad, always out of reach, too swift for pursuit—a vector (a force with direction), marking the primacy of movement in smooth space. "The edginess of wild place-worlds is due to the radical becoming of nature itself" (Casey 139), which the Creature embodies, and these encounters reveal. While Shelley's novel conveys the radical becoming of wilderness in its spell-binding Alpine and Arctic passages, as we have seen, its acumen comes across perhaps most acutely in the Creature's dynamic radicality. This is reflected in the novel's ethos and narrative but also in its language. The passages above, pivoted on the verb and other signifiers of speedy disappearance, showcase the primacy of movement in the Creature's nomadic characteristics.

The Creature subsists in wilderness and dictates the nomadic in the plotline, leading Victor and the reader deep into uncharted smooth space. If nomads are "vectors of deterritorialization," adding wilderness to wilderness "by a series of local operations whose orientation and direction endlessly vary" (DG 382), the Creature deterritorializes the narrative:

> the mist cleared away, and we beheld, stretched out in every direction, vast and irregular plains of ice, which seemed to have no end . . . [A] strange sight suddenly attracted our attention . . . We perceived a low carriage, fixed on a sledge and drawn by dogs, pass on towards the north, at the distance of half a mile: a being which had the shape of a man, but apparently of gigantic stature, sat in the sledge . . . We watched the rapid progress of the traveller with our telescopes, until he was lost among the distant inequalities of the ice. (13)

This first baffling appearance of the Creature vanishing into the Arctic wilderness marks the gradual transformation of the entire text—its narrative structure and discursive energy—into a smooth-space wilderness, an "acentered, directional" "multiplicity" (DG 484). The reading process, a piecing together of the multiplying fractured and embedded narratives, each "acentered" and "directional" in its own right, assumes the dynamics of nomadic movement. As Nail explains, since "the figure of the nomad . . . has been abandoned by the centripetal force of territorialization, its point of counterpower or pressure is precisely its attempt to break into the territory via the raid" (134). The Creature's raid of the local Arctic winter camp is a case in point here, but much more crucially, the Creature's nomadic force invades the text by his embedded narrative, expounded during the pivotal meeting on the glacier and reported in the Arctic by Victor relating it to Walton, splintering into yet further embedded narratives. The diegetic territory is thus invaded by a series of hypodiegetic tales, and later invaded further still by the Creature's actual diegetic presence in the Arctic finale. One of the masterstrokes of Shelley's novel is its narrative layering, the embedded narratives disrupting rather than edifying the already fragmented diegetic flow of Walton's epistolary frame. In this sense, the novel's metaleptic transgressions correspond to the characteristics of the nomadic.

Frankenstein's elaborate conceptualization of wilderness is essentially transgressive—on discursive, philosophical, and ethical as well as narrative levels. As Wolfgang Iser suggests, "it is only through inevitable omissions that a story will gain its dynamism." A text is

> full of unexpected twists and turns, and frustrations of expectations . . . Thus whenever the flow is interrupted and we are led off in unexpected directions, the opportunity is given to us to bring into play our own faculty for establishing connections—for filling in the gaps left by the text itself. (Iser 280)

This marks the dynamic interaction between the reader's predisposition to striate, "establishing connections" to construct an ultimately linear meaning-oriented trajectory, and the wilderness of textual territory. *Frankenstein*'s essentially nomadic narrative structure resists linearity and closure. In the nomadic, "every point is a relay and exists only as a relay" (DG 380), and the novel's embedded narratives function as relays of meaning. The reading of an experimental text such as *Frankenstein* challenges the reader's response mechanism and intensifies participation in the performance of the text. It is a living process of essential transgression, because "movement, as a continuous flow, is always both/and: it is an inclusive disjunction" (Nail 26). Among the plethora of the radically new in Shelley's novel, her intricate, multi-layered treatment of wilderness and her bold critique of Enlightenment and Romantic fallacies relating to it, many of which, regrettably, still determine our present perceptions of wild nature, are perhaps more relevant to our own troubled consciousness of the Anthropocene than ever before.

Works cited

Barrow, John. *A Chronological History of Voyages into the Arctic Regions*. John Murray, 1818.

Bate, Jonathan. *Song of the Earth*. Picador, 2001.

Cantor, Paul. *Creature and Creator: Myth-Making and English Romanticism*. Cambridge UP, 1984.

Carroll, Siobhan. *An Empire of Air and Water: Uncolonizable Space in the British Imagination 1750–1850*. U of Pennsylvania P, 2015.

Casey, Edward S. *The World on Edge*. Indiana UP, 2017.

Cavell, Janice. "The Sea of Ice and the Icy Sea: The Arctic Frame of *Frankenstein*." *Arctic*, vol. 70, no. 3, September 2017, pp. 295–307.

Coxe, William. *Travels in Switzerland*, 2 vols. James Decker, 1802.

Craciun, Adriana. *Writing Arctic Disaster: Authorship and Exploration*. Cambridge UP, 2016.

Davidson, Peter. *Distance and Memory*. Carcanet, 2013.

———. *The Idea of North*. Reaktion Books, 2007.

Deleuze, Gilles and Félix Guattari. *A Thousand Plateaus: Capitalism and Schizophrenia*. Translated by Brian Massumi. Continuum, 1987.

Duffy, Cian. *The Landscapes of the Sublime 1700–1830: Classic Ground*. Palgrave, 2013.

Foucault, Michel. "A Preface to Transgression." *Language, Counter-Memory, Practice: Selected Essays and Interviews*, edited by Donald F. Bouchard. Translated by D. F. Bouchard and Sherry Simon. Cornell UP, 1977, pp. 29–52.

Fulford, Tim, Debbie Lee, and Peter J. Kitson. *Literature, Science and Exploration in the Romantic Era*. Cambridge UP, 2004.

Griffin, Andrew. "Fire and Ice in Frankenstein." *The Endurance of "Frankenstein": Essays on Mary Shelley's Novel*. Ed. George Levine and U. C. Knoepflmacher. U of California P, 1979, pp. 49–73.

Heymans, Peter. *Animality in British Romanticism: The Aesthetics of Species*. Routledge, 2012.

Iser, Wolfgang. *The Implied Reader: Patterns of Communication in Prose Fiction from Bunyan to Beckett*. Johns Hopkins UP, 1974.

Lopez, Barry. *Arctic Dreams: Imagination and Desire in a Northern Landscape*. Picador, 1986.

Lörke, Melanie Maria. *Liminal Semiotics: Boundary Phenomena in Romanticism*. Akademie Verlag, 2013.

McCannon, John. *A History of the Arctic: Nature, Exploration and Exploitation*. Reaktion Books, 2012.

Nail, Thomas. *The Figure of the Migrant*. Stanford UP, 2015.

Oelschlaeger, Max. *The Idea of Wilderness: From Prehistory to the Age of Ecology*. Yale UP, 1991.

Shelley, Mary. *Frankenstein* [1818]. Ed. J. Paul Hunter. 2nd ed. W. W. Norton, 2012.

———. *Frankenstein* [1831]. Ed. M. K. Joseph. Oxford UP, 1998.

Shelley, Mary and P. B. Shelley. *History of a Six Weeks' Tour*. T. Hookham, Jun. and C. and J. Ollier, 1817.

Spufford, Francis. *I May Be Some Time: Ice and the English Imagination*. Faber and Faber, 1996.

Wilson, Eric G. *The Spiritual History of Ice: Romanticism, Science and the Imagination*. Palgrave, 2003.

Index

Note: *Italic* page numbers refer to figures and page numbers followed by "n" denote endnotes.